AF540316

MICROBIOLOGY

MICROBIOLOGY

By

MANJU YADAV

Lecturer

Department of Zoology

M.M.H. College

Ghaziabad (U.P.)

(India)

DISCOVERY PUBLISHING HOUSE

NEW DELHI-110002

First Published – 2007

Reprinted – 2023

ISBN: 978-81-7141-731-5

© Author

Microbiology

Published by:

DISCOVERY PUBLISHING HOUSE

4383/4B, Ansari Road, Darya Ganj

New Delhi-110 002 (India)

Phone: +91-11-23279245; 23253475; 43596065

+91 9811179893 / +91 9871656464

E-mail: discoverybooksindia@gmail.com

orderdphbooks@gmail.com

namitwasan9@gmail.com

web: www.discoverypublishinggroup.com

Printed at:

Infinity Imaging Systems

Delhi

Preface

The present title "Microbiology" has been carefully compiled and edited to meet the long felt needs of increasing number of students and researchers who have to deal with the pharmacy, agriculture, environmental science, waste disposal industry, biotechnology and also medicine. An attempt has been made to introduce the readers to several area of microbiology and provide insight into basic concepts of biology of micro organism. The material presented in the following pages will afford the students in the subject a comprehensive view and appropriate background material for focussed studies. The way of presentation is very clear and lucid, which can be easily followed by the students. It is our earnest hope that this bok will be of great value to all our students.

Author

Contents

1

History of Microbiology

One can't overemphasize the importance of microbiology. Society benefits from micro–organisms in many ways. They are necessary for the production of bread, cheese, beer, antibiotics, vaccines, vitamins, enzymes, and many other important product. Indeed, modern biotechnology rests upon a micro–biological foundation. Micro–organisms are indispensable components of our eco-system. They make possible the cycles of carbon, oxygen, nitrogen, and sulfur that take place in terrestrial and aquatic systems. They also are a source of nutrients at the base of all ecological food chains and webs.

Of course micro–organisms also have harmed humans and disrupted society over the millenia. Microbial diseases undoubtedly played a major role in historical events such as the decline of the Roman Empire and the conquest of the New World. In 1347 plague or black death struck Europe with brutal force. By 1351, only four years later, the plague had killed 1/3 of the population (about 25 million people). Over the next 80 years, the disease struck again and again, eventually wiping out 75% of the European population. Some historians believe that this disaster changed European culture and prepared the way for the Renaissance. Today the struggle by microbiologist and others against killers like AIDS and malaria continues.

In this introductory chapter the historical development of the science of microbiology is described, and its relationship to medicine and other areas of biology is considered. The nature of the microbial world is then surveyed to provide a general idea of the organisms and agents that microbiologist study. Finally, the scope and relevance of modern microbiology are discussed.

Microbiology often has been defined as the study of organisms and agents too small to be seen clearly by the unaided eye that is, the study of *micro–organisms*. Because objects less than about one millimeter in diameter cannot be seen clearly and must be examined with a microscope. Microbiology is concerned primarily with organisms and agents this small and smaller. Its subjects are viruses, bacteria, many algae and fungi, and protozoa. Yet other members of these groups, particularly some of the algae and fungi, are larger and quite visible. For example, bread molds and filamentous algae are studied by microbiologist, yet are visible to the naked eye. A bacterium that is visible without a microscope, *Epulopiscium*, also has been discovered. The difficulty in setting the boundaries of microbiology led Roger Stanier to suggest that the field be defined not only in terms of the size of its subjects but also in terms of its techniques. A microbiologist usually first isolates a specific micro-organism from a population and then cultures it. Thus microbiology employs techniques such as sterilization and the use of culture media that are necessary for successful isolation and growth of micro–organisms.

Early Development of Microbiology

Leeuwenhoek

In 1963, in Delft, Holland, a successful linen merchant by the name of Anton van Leeuwenhoek first observed and consequently introduced man to the mysterious and exciting world of micro-organisms. For fifty years, until his death at the age of 90 in 1723, Leeuwenhoek continued to make countless observations with the aid of small, simple microscopes. Even though compound microscopes had been developed long before (1590), Leeuwenhoek found his opitcal device more suitable for observing specimens with transmitted light. The magnifying power of his early instruments ranged from approximately 50 to 300 times the diameter of a particular specimen.

Anton van Leeuwenheok's position in the development of microbiology had been firmly established, not because he constructed microscopes with improved lenses, but because of his remarkable observations and descriptions of microscopic forms of life. Due to his unending curiosity about the world around him, he spent hours upon hours examining specimens that he collected from lakes, rain barrels, and even from his own teeth and those of other people. Among his first observations can be found descriptions of protozoa, the basic shapes of bacteria, yeasts, and algae. Because of the importance of these observations,. Leeuwenhoek is considered by many to be the *father of Bacteriology and the Father of Protozoology.*

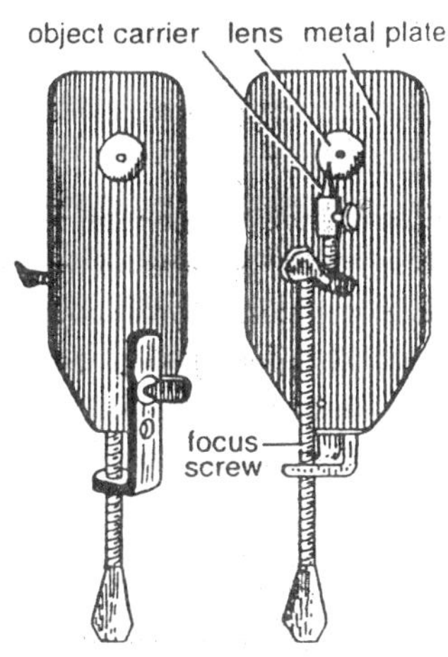

Fig. 1.1 Microscope of Leeuwenhoek

Leeuwenhoek's discoveries went beyond the "microbial" world for he made numerous other contributions of biological significance. For example, he provided confirmatory evidence for William Harvey's theory of blood criculation by constructing an aquatic microscope ("alkijer"). This instrument enabled him and others to observe the flow of erythrocytes through the capillaries of a fish's caudal fin. Some of Leeuwenheok's other observations included the demonstration of muscle fibres striation, nuclei of fish erythrocytes sand the myelin sheath of nerve fibres. No wonder such discoveries were provocative to investigators and inquisitive amateurs everywhere. Unfortunately, the investigations of micro-organisms were neglected for some time after Leeuwenhoek's death. This was partly due to difficulties encountered in the construction of better microscopes, and partly to the fact that many still consider micro-organisms to be nothing more than little oddities.

Pasteur

Except for Leeuwenhoek, French and German scientists dominated the field of early microbiology. One of the first scientists to impart a true biological function to microbes was Louis Pasteur, truly one of the major figures in the development of biology and medicine. A chemist and physicist, Pasteur was born on December 27, 1882, in the little French village of Dole. The discoveries of this man were destined not only to provide additions to existing knowledge, but to bring to light dramatic new concepts and approaches to age-old problems. Pasteur's contributions have been responsible for new and more effective measures for the prevention of disease, the improvement of health in general, and the understanding of basic aspects of microbial life. The latter includes such things as the processes of fermentation, pasteurization, and the development of effective vaccines against such dreaded diseases as rabies and anthrax.

The Germ Theory of Fermentation

Non-vitalist vs Vitalist

Fermentation is a process in which alcohols and organic acids such as vinegar or lactic acid are formed from sugar containing fluids. The results of fermentation reactions including the souring of milk and the preparation of alcoholic beverages, have been observed and used by people all over the world throughout history. Yet, despite the recorded descriptions of micro-organism by Leeuwenhoek, the true biological basis of fermentation was not formulated until well into the nineteenth century. Basically, two major schools of thought evolved as explanations for these processes, namely, the non-vital (nonbiological) and the vital (biological) theories.

According to the non-vital view, the yeasts seen in fermenting materials were considered to be the by-products rather than the causes of fermentation. During the period from approximately 1839 to 1859, supporters of the non-vital theory, including the three influential chemists Berzelius, Liebig, and

Wohler, maintained that essential, unstable chemical entities called *ferments* produced the reactions in question as catalysts, enzymes, or simply activators of chemical reactions. These unstable substances came into being as a consequence of the action of air upon sugar-containing fluids. The resulting ferments transmitted their property of instability to sugar molecules, which in turn decomposed, thus forming the products of fermentation. Liebig used as support for this non-vital theory his observations showing the absence of yeasts in acetic and lactic acid fermentations. Unfortunately, the non-vitalists neglected to consider that other micro–organisms could be the producers of these "essential ferments". Several years later these reactions were shown to be caused by bacteria.

The basis of the biological theory of fermentation was established independently and almost simultaneously during the years of 1836 and 1837 by three scientists the German algologist Kutzing, the French physicist Cagnaird Latour, and the German physiologist-pathologist Schwann. Schwann clearly demonstrated the role of yeasts in alcoholic fermentation and their sensitivity to heat and chemical agents. Threating these micro–organisms by such means stopped all fermentative activity. In addition, Schwann described the asexual form of reproduction (budding) of yeast, and showed that this "sugar fungus" (Saccharomyces cervisiae) was needed in large numbers for the fermentation reaction to proceed. Needless to say, these obsevations were not readily accepted by the non-vitalists. Thus the stage was set for a bitter controversy which was not settled until Louis Pasteur provided the crucial experimental proof for the microbial nature of fermentation in 1857.

Pasteur's Contributions

Pasteur's studies involving fermentation occupied a major portion of his scientific career, extending from approximately 1854 to 1876. His entry into this field came about through a request for help in finding the basis for the souring and spoilage of beer and of wines. Pasteur was a professor at the University of Lille, France, in a city where the production of wine and beer was a very important industry. Pasteur found that the problem was caused by a type of fermentation process other than one involving normal yeasts. In short, sugar was converted to lactic acid rather than alcohol, by lactic acid fermentation. The microscopic examination of sediments from wine vats in which the undesirable reactions occurred showed the presence of micro–organisms which eventually were recognized as bacteria and unwanted ("wild") yeasts. The classification of such organisms was difficult since the so-called "formal rules" of taxonomy was not yet developed. The obvious solution to the problem was to find a way to eliminate and/or destoy these microbes without altering the quality of the alcoholic beverages.

On further experimentation, Pasteur came to realize that wines could be heated and held for some time without spoiling at a temperature intermediate between 50° and 60°C (122° and 139° F). This procedure came to be known as *pasteurization*. It was subsequently applied to a variety of products, the best known of which is milk. Contrary to popular opinion, pasteurization was not applied to milk first. Today routine pasteurization can be performed by heating milk or other beverages at 63°C (145° F) for 30 minutes. This exposure, generally speaking, is sufficient to kill pathogenic (disease-causing) micro–organisms.

In investigating many other fermentative processes, Pasteur discovered that:

1. Each type of chemical fermentation as defined by the particular organic end product or products formed, is associated with a specific microbial type.
2. Specific environmental conditions are necessary for the development of each micro–organisms, such as a definite degree of acidity or alkalinity.

While studying a particular fermentation, Pasteur discovered micro–organisms that could live only in the absence of free oxygen, and named them *anaerobes*. He became aware of this fundamental biological phenomenon during microscopic examination of bacteria causing butyric acid fermentation. While these organisms normally are motile, he noticed that certain ones in a preparation, specifically those in close contact with air, ceased to move, i.e., became immotile. Pasteur quickly realized the possibility that air might exert an inhibitory effect on these bacteria, and confirmed his suspicion by introducing a stream of air into fermentating systems. The effects was obvious.

The process was either totally stopped or slowed down considerably. The terms "aerobic" and "anaerobic" were coined by Pasteur to distinguish between microbes capable of living in the presence and absence of free oxygen, respectively.

The contributions of Pasteur were to have a pronounced effect on the control of micro–organisms and demonstrated the relationship of numerous organisms to "disease." As pointed out by Stanier, Doudoroff, and Adelberg, Pasteur referred to the spoilage processes of beer and wine as "diseases." Perhaps in his own mind the association between micro–organisms and the infectious diseases of animals and plants was firmly fixed.

The Spontaneous Generation or Abiogenesis Controversy

The spontaneous appearance of life from inanimate of decomposing organic matter was a belief commonly held by man since at least 346 B. C. This concept appears to have been proposed by the Greek scholar Aristotle and perpetuated by his students and numerous others well into the nineteenth century. The "Aristotelian doctrine of spontaneous generation" met little opposition, because man constantly saw what he thought to be examples of the process. The appearance of snakes, frogs, and related forms of life from the mud of river banks, and the development of maggots in, and the emergence of flies from, decaying food supported the unquestioning acceptance of this doctrine. It was thought that lower forms of life could not originate by any other means. Thus, before any relationship of micro–organisms to various natural processes, including disease, could be shown, the concept of spontaneous generation would have to be disproved.

Francesco Redi

Among the first notable demonstrations that the doctrine of spontaneous generation did not apply to highly organized animals was made by the Italian naturalist Franscesco Redi. In approximately 1665 he showed that maggots did not emerge spontaneously (*de nova*) from putrefied (*decayed*) meat. Redi put meat into three separate containers or flasks. One of these was stoppered, another was left uncovered, and the third was covered with a gauze or veil-like material. Naturally the meat readily putrefied and attracted flies. Redi made the following observations:

1. The stoppered container showed no evidence of any form of flies, whether larvae or adults;
2. Flies laid their eggs on the meat in the uncovered container. And, within a short period of time, the larval stages or maggots and newly emerging adult flies appeared;
3. Although no maggots were present in the meat in the gauze-covered flask, they did appear on the covering. Apparently the aroma of the putrefying meat attracted the flies, but because they were unable to gain access to it, they simply laid their eggs on the gauze.

Thus Redi dealt a crushing blow to the myth that flies were spontaneously generated from meat. However, even with his evidence in hand against the doctrine of spontaneous generation, Redi still believed that insects in a type of plant tumor called *galls* arose spontaneously.

The "War of Infusions"

Leeuwenhoek's discovery of bacteria so-called animalcules-revived the arguments for the occurrence of spontaneous generation but on a microscopic level. Although he felt that his newly discovered forms came from the surrounding air, Leeuwenhoek performed no systematic study to prove it. Another view shortly developed. Many individuals believed that inanimate animal or vegetable matter held a "vital" or life-generating force which could be transformed so as to give rise to animalcules. The appearance of bacteria and protozoa, the so-called *Infusoria*, in boiled preparations of hay or meat after standing was considered as ample supportive proof. This type of solution is known as an *infusion*. In 1711 Louis Joblot, a French contemporary of Leeuwenhoek's, carried out a study which showed this was not the case. Basically, he observed that infusions stoppered tightly immediately after boiling remained free of micro–organisms. However, if such boiled and stoppered preparations were later opened and exposed to the air of the environment, animalcules soon appeared. Joblot's findings were challenged, and thus the "war of infusions" began.

Johan Needham

This Roman Catholic priest in 1749 reported the results of his experiment which he felt proved that bacteria arose spontaneously in an environment where no such living forms existed before. His studies primarily consisted of tightly corking flasks containing boiled mutton broth and periodically observing them for cloudiness as an indication of microbial growth. While some containers remained clear, most of them eventually became cloudy or turbid. Upon examining a few drops of fluid from these preparations with the aid of a microscope, he found them teeming with micro–organisms. Since boiling was known to destroy micro–organisms as well as any other living cells, Needham felt justified in concluding that his experiments clearly demonstrated the existence of spontaneous generation. He postulated that the organic matter in his flasks possessed a "vital or vegetative force" which could confer the properties of life on the non-living elements present.

Lazzaro Spallanzani

Some years later, in 1765, the Abbe Lazzaro Spallanzani, an Italian naturalist, re investigated Needham's findings and conclusions. He primarily questioned the heating procedure used by his predecessor. Spallanzani found that heating hermetically sealed flasks containing any one of several types of organic matter for one hour did not cause them to become cloudy within a reasonable length of time. This experiment was repeated several times with the same results occuring every time. Needharm argued that the prolonged boiling procedure destroyed the life- rendering "vegetative force." Spallanzani responded to this argument by taking heated, closed flasks and breaking the seal allowing exposure to air. Within a short time the contents of these flasks became turbid, thus showing that the long heated organic matter still was capable of supporting life.

Spallanzani's experiment appeared to have dealt the doctrine of spontaneous generation a crucial blow. However, the effect was short-lived. This was largely because of the discovery of oxygen by Joseph Priestly, a Unitarian Minister, and the demonstration of its importance to life by Lavoisier in 1775. The arguments for spontaneous generations began once again, as Spallanzani's findings were criticized from the standpoint that sufficient air was not present in his sealed flasks to support microbial growth.

T. Schwann and F. Schulze

Additional experiments now were necessary to show that bacterial growth in nutrient-containing flasks was brought about through an exposure to air containing these organisms, and that the result was not a case of spontaneous generation. Theodor Schwann in 1836 arranged separate series of flasks, one of which held an infusion of some type, to prove this point. Air entering one such experimental system was passed through a tube device heated red-hot. Following this exposure, the air was introduced into the flask containing the nutrient material. Another system received untreated air and thus served as the experimental control. Soon growth developed in the control, while the system receiving heated air remained sterile (free of any living organisms). Similar experiments were performed by Franz Schulze in 1836. However, his approach involved the use of the chemical agents sulfuric acid and potassium hydroxide. Air was allowed to enter flasks with nutrients only after it had passed through these compounds.

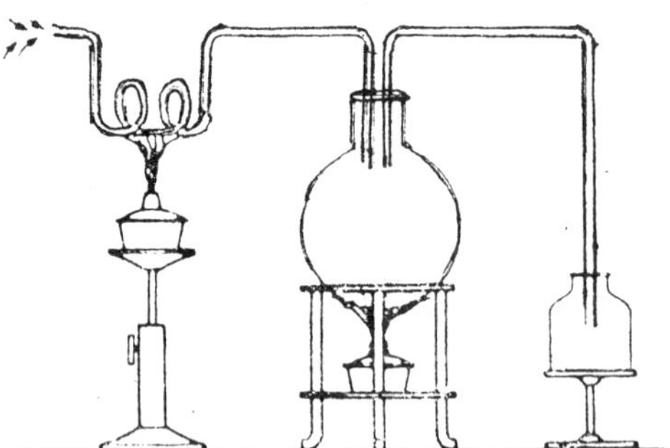

Fig. 1.2 A diagrammatic representation of Schwann's experimental system to disprove spontaneous generation.

Schroder and von Dusch

Upon hearing the results of these experiments, the proponents of spontaneous generation contended that the drastic treatments to which the air systems were subjected in the studies of Schwann and Schulze destroyed all possible "life rendering power," This obviously would prevent life from being spontaneously generated. This objection was countered in 1854 by Schroder and von Dusch, who introduced the practice of using cotton plugs for bacteriological culture flasks and tubes. While their experimental design was similar to that of Schulze and Schwann, these scientist did not treat air in any way other than simply filtering it thorough cotton wool which had been previously baked in an oven.

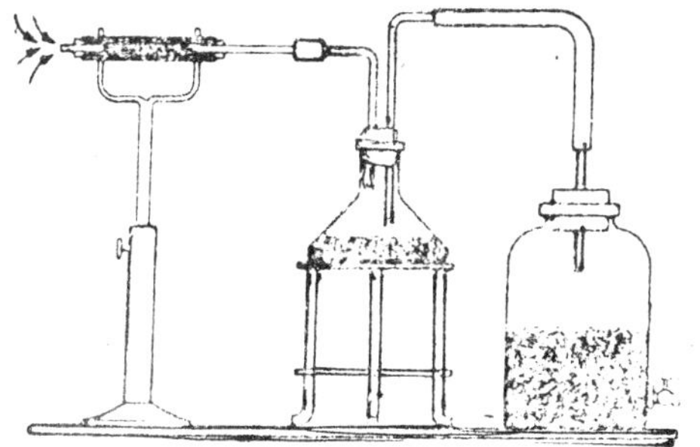

Fig. 1.3 A diagrammatic representation of the experimental system used by Schroder and von Dusch in 1854 to demonstrate the removal of living organism.

The results of these experiments were the same. Flasks which received "filtered" air showed no signs of growth. While those systems exposed to unfiltered air clearly demonstrated the presence of micro–organisms. These studies confirmed that the treatment of air with chemicals or with heat in reality was unrelated to the development of growth in nutrient-containing flasks. Moreover, these findings demonstrated not only the sensitivities of living forms in air to chemicals and heat, but the fact that they could be removed from air by filtration through cotton wool. Pasteur later demonstrated the presence of bacteria in the cotton wool used for filtration, showing that these organisms were trapped in the material.

The Final Blows to Spontaneous Generation.

The Contributions of Pasteur and Tyndall

Although these various experiments might seem conclusive, the issue was far from settled. This was evident from the 1859 and later writing of one of Pasteur's principal antagonists, Felix Pouchet. Pouchet claimed to have carried out experiments, which conclusively showed that microbial growth could occur in the "absence of atmospheric contamination." About this time the studies of Pasteur were becoming public knowledge and several other scientists began to recognize the roles of micro-organisms in fermentation and putrefaction processes. However, the acceptance of his views on the biological function of microbes was threatened by the claims of Pouchet and other supporters of spontaneous generation. Irritated by these arguments, Pasteur was determined to disprove spontaneous generation once and for all.

Slightly altering the procedure of Schroder and von Dusch, Pasteur passed large volumes of air through a tube which held a plug of guncotton serving as a filter. A portion of this material was then dissolved in an alcohol-ether mixture and the sediment remaining was examined microscopically. Pasteur found small round or oval bodies which resembled the spores (reproductive structures) of plants. To show that the guncotton not only stopped the passage of micro–organisms, but was heavily laden with these forms of life, Pasteur simply added a little of the used filter to sterile meat infusions. Soon microbial growth appeared. Thus Pasteur confirmed how microbes gained access to and also how they could be prevented from entering fermentable and related types of nutrients.

Despite his obvious success, Pasteur was not fully satisfied. He then performed what has been referred to by Stanier, Doudoroff, and Adelberg as "perhaps his most elegant experiments on the subject" to show that air lacking in microbes could not create life from organic infusions. Special swan-neck or gooseneck, flasks were made to which liquid nutrient media was added. No plugs of any type were used to prevent the passage of micro–organisms into these systems. The flasks and their contents were first sterilized by boiling. Despite the fact that these systems were open to the external environment, growth did not develop. Because of the length and the bend of the flasks' gooseneck, micro–organisms present in the air could not be transported into the flask proper. However, if the top of a system were broken off, or if a flask were tilted so that the sterile liquid nutrient ran into the exposed part of the neck and then returned, microbes soon appeared in the fluids.

Pasteur also demonstrated that the distribution of micro–organisms is not uniform in the atmosphere. During a summer holiday, he took several hermetically sealed flasks containing sterile nutrient fluids to many localities in France. A certain number of these flasks were opened at each location, exposed to the environment and then quickly sealed. Flasks that were exposed to the atmospheres of deep wine cellars or mountain air

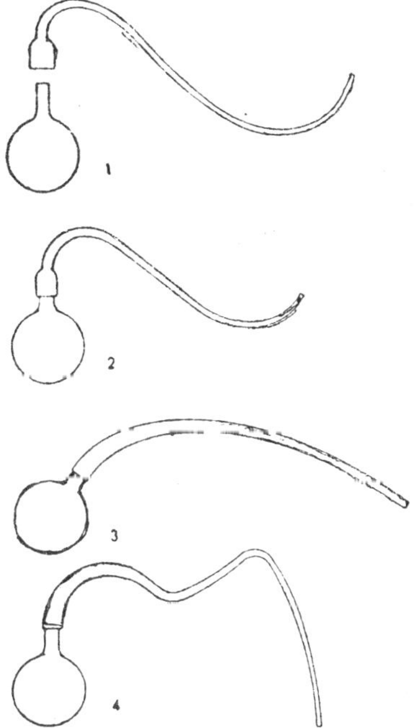

Fig. 1.4 "Swan-neck" flasks used by Pasteur in his experiments to disprove the doctrine of spontaneous generation.

in the Alps mostly remained free of micro-organisms. However, the experiments carried out in the streets of Paris, or on the road to Pasteur's home town, Dole, produced several contaminated flasks. Thus Pasteur showed in a semi-quantitative fashion that micro–organisms, although present in the atmosphere, are not evenly distributed.

Most authorities agree that the "final blow" to spontaneous generation was given by British physicist Johan Tyndall in 1877. During the course of his studies concerned with the optical properties of atmospheric dust, he observed that a beam of light passing through air lacking dust particles could not be seen. On the other hand, Tyndall found that a light beam passed through a "dust-laden" environment was readily visible. Moreover, the dust particles within such an atmosphere could also be seen. Aware of Pasteur's conclusions regarding the presence of micro–organisms on dust, and the greater likelihood of microbial contamination in a dusty environment. Tyndall devised a system to determine if air lacking dust particles ("optically empty air") contained micro–organisms. He built a specialized culture chamber equipped with vents through which bacteria could not enter, lateral windows, and test tubes in racks. In addition, the sides of this box were coated with glycerol in order to trap the dust particles in the chamber which sooner or later would come to settle on the surfaces. When the chamber was found to be optically devoid of floating matter, as determined by shining a beam of light through its lateral windows, the test tubes were filled with a broth medium, which was then sterilized by placing the bottoms of these tubes in a pan of boiling brine. Tyndall observed that the broth remained sterile even though it was in direct contact with the air of the chamber. When dust-laden air was introduced, microbial growth appeared after a brief incubation period. Thus, with his specialized chamber and techniques, Tyndall demonstrated that bacterial life occurred in sterile broth only after it was introduced from an outside source.

During the course of his studies, Tyndall also became aware of incredibly resistant bacterial structures. These forms, which are now known as spores, were named and independently demonstrated in 1877 by the German botanist-bacteriologist

Ferdinand Cohn. While attempting to repeat his experiments with dust-laden environments, Tyndall found he was unable to obtain similar results after a bale of hay (used in broth preparations) was brought into his laboratory. This situation arose largely because of the presence of spore-forming bacteria in the hay which interfered with the sterilization of nutrients in test tubes. Only when tests were performed in different rooms could the results of his original experiments be duplicated. The major difficulty was eliminated by boiling nutrient-containing solutions for short periods of time on each of three successive days, and incubating them between sterilizations at favourable temperatures to allow microbial growth to occur. This process of intermittent sterilization subsequently became known as *tyndallization*. Today, spores are destroyed in bacteriology laboratories by more rapid means, with the apparatus called the *Autoclave*. This device incorporates steam under pressure, usually at a temperature of 121.5°C.

Antiseptic Surgery

The introduction of anesthetics into surgery and obstetrics during the 1840s contributed greatly to the development of efficient surgical techniques, enabling surgeons to perform complex and lengthy operations which previously were not feasible to undertake. Unfortunately, the incidence of wound infections from surgical procedures increased and quite often resulted in the death of patients. Confronted with this problem and the desire to prevent wound infections, the young English surgeon Josph Lister, undertook the task of combating them.

Impressed with Pasteur's studies showing the involvement of micro–organisms in fermentation, putrefaction, and spontaneous generation, Lister reason that the very basis of surgical infection, sepsis, might be microbial in nature. Consequently, he devised a series of procedures designed to prevent the access of micro–organisms to wounds. This system came to be known as antiseptic surgery. It included the heat sterilization of instruments and the application of carbolic acid (phenol) to wounds by means of dressings. Lister's procedures were critically received at first, but ultimately proved to be an effective means of preventing surgical sepsis, thus establishing modern surgery. Although Lister was not aware of the exact nature of the micro–organisms involved, antiseptic surgery did provide an indirect source of evidence in support of the germ theory of disease.

The Germ Theory of Disease

In developed countries today, most of the major pestilences, including cholera, plague, smallpox, typhoid, typhus, and yellow fever, which were responsible for the deaths of millions of people in the past, are controlled by means of prophylactic Vaccination, environmental sanitation and the destruction of arthropod vectors ("bugs" such as fleas, lice mosquitoes and ticks that serve to transmit specific infections). Unfortunately, in so-called "undeveloped" countries, many of these diseases still take a heavy toll of life, and also cause significant disability in the population. Needless to say, the discovery of the causative agent of a particular disease is a major stepping stone in its control. Individuals such as Robert Koch, Louis Pasteur, and others who established the specific relationship between a disease agent and a disease state, and who developed methods for the control of infections, deserve to be heralded as major contributors to the well-being of mankind.

From the earliest times, diseases were associated with natural phenomena, such as earthquakes and floods, mysterious and supernatural forces, and poisonous vapours called miasmas. Although ancient Greek and Roman physicians suspected that certain types of diseases were caused by invisible, minute agents, no direct proof for this view was found until the nineteenth century. The concept of contagion the discovery of how infectious diseases were spread from a diseased individual to others preceded the demonstration of the existence of pathogenic agents by many centuries.

Fungi were the first micro–organisms shown to be pathogenic. Agostino Bassi proved experimentally that an agent of this type caused an infection in silkworms. This discovery was followed in 1839 by the first isolation of a fungus from a human skin disease by Schonlein. As later chapters will show, many fungi are recognized as a serious threat not only to man, but to the plants and other forms of animal life in his environment.

Protozoa also were among the first micro–organisms shown to have an association with disease. This relationship is credited to Pasteur, who in 1865 discovered that an infection of silkworms, which were vital to the silk industry in Europe, was "protozoan" in nature. The disease was called pebrine.

Koch's Postulates

The direct demonstration of an etiologic role of bacteria as agents of an infectious disease was given by Koch in 1876, and confirmed by Pasteur and Joubert. The organism used was *Bacillus anthracis*, the cause of anthrax. Although rod-shaped structures had been observed by Davaine several years earlier in the blood and organs of sheep dying of anthrax, there was no clear-cut proof at that time that these bodies were the cause of anthrax.

Koch established a definite sequence of experimental steps or rules with which the causal relationships between a specific organism and a disease state could be proved beyond a doubt. Although this procedure is known as "Koch's Postulates," it is important to note that Jacob Henle, a German scientist, offered the theoretical basis for the demonstration of the germ theory of disease in 1840. In showing the causal relationship of B. *Anthracis* to anthrax, Koch first had to isolate the organism form a case of the disease and ultimately obtain similar cultures from laboratory-inoculated animals exhibiting symptoms of the infection.

The steps of Koch's Postulates can be generalized as follows:

1. The causative agent must be found in every case of the diseases;
2. This micro–organism must be isolated from the infected host, or patient, and grown in a pure culture containing no other kinds of micro–organism (Koch utilized the aqueous humor of the eyes of cattle for this purpose);
3. A pure culture of the suspected agent must reproduce the specific disease after its inoculation into a new normal, healthy, susceptible animal;
4. The same micro–organism must be recovered again from the experimentally infected host.

With relatively few exceptions, the causal relationship of pathogenic bacteria to a particular disease state has been shown according to the dictates of Koch's Postulates. One notable exception is the causative agent of human leprosy, *Mycobacterium leprae,* for which man is the only natural known host. Attempts to reproduce the disease with organisms isolated from actual human infections have met with repeated failures and only a few successes under experimental conditions.

Rivers' Postulates

At the time Koch formulated his system, true viral pathogens were unknown. The need for criteria to demonstrate their relationship to diseases became apparent shortly after their discovery. In 1937, Rivers created a group of rules similar to those of Koch for this purpose. River's Postulates briefly are as follow's applicable to both animal and plant viruses:

1. The viral agent must be either demonstrated in the host's body fluids, e.g., blood and spinal fluid, or plant sap, at the time of the disease, or present in the cells of the host showing specific lesions;
2. Filtrates obtained from the tissues or body fluids of an infected host must produce the specific disease in a suitable healthy animal or plant or provide evidence of infection in the form of antibodies against the viral agent. It is important to note that all filtrates used for inoculations must be free of any bacteria or other microscopic cultivable micro–organisms;
3. Similar filtrate material from such newly infected animals or plants must in turn be capable of transmitting the specific disease in question to other hosts.

The Microbiologist

Today, elaborate equipment and procedures play an important role in microbiology, but the most important instrument must still be the mind of man. Scientific investigators must exercise great care in not allowing their views and feelings to influence them. This becomes extremely difficult when the findings of a study are contrary to the beliefs of the person conducting the research experiment. Nevertheless, all scientists are bound by principles of integrity. Failure to observe this responsibility may have dire consequences, especially in the eyes of the scientist's fellow workers.

Public recognition of the contributions by scientists takes many from. The Nobel Prize is one example. These awards were established by Alfred Bernhard Nobel, the inventor of dynamite. From his discovery Nobel became an extremely wealthy industrial magnate. In 1896 he died, leaving most of his accumulated fortune for use in the awarding of the Nobel Prizes. Noble prizes are awarded to individuals "who, during the preceding year, shall have conferred the greatest benefit on mankind." The categories for consideration include chemistry, economics, literature, medicine or physiology, peace, and physics.

Recognising the Relationship between Micro–organisms and Disease

Our earliest ancestors had to contend with disease. Major outbreaks of disease ran rampant through populated areas. The Paleological records leave no doubt therefore that most of the known organic and microbial disorders of man and animals are extremely ancient. But it is also certain that the comparative prevalence and severity of various diseases have changed greatly in the course of historical times.... Influenza, yellow fever, smallpox, typhus, and cholera likewise conjure up the thought of visitations descending on mankind as the curse of some avenging deity. Other epidemics are now forgotten, but caused great terror in their days.

The various epidemics of plague and other diseases constituted great destructive cataclysms. During the Justinian era plague killed one half to two thirds of the inhabitants of certain cities of the Roman world. In Europe leprosy was prevalent in the fourteenth century, plague in the fifteenth, syphilis in the sixteenth, smallpox in the seventeenth and eighteenth centuries, scarlet fever, measles, and tuberculosis in the nineteenth century.

When, in the mid-fourteenth century more than 25 million people died in medieval Europe during a severe outbreak of plague, the populace turned to such unscientific and unrewarding practices as flagellation— beating themselves— to drive out the force causing "black death." A more relevant method of dealing with plague was that of removing individuals suspected of having the diseases from contact with the general population. This practice dates back to Biblical times. At the height of the fourteenth century, when plague was epidemic, sea voyagers coming into Sicily had to wait 40 days before entering the city— a practice known as quarantine.

Two centuries after the major outbreaks of plague in Europe, Girolamo Fracastoro on Verona, a contemporary of Copernicus, published a work on contagious diseases and their treatment. *De-Contagione* (1546) was largely philosophical and did not recognize the true nature of micro–organisms. Nevertheless, Fracastoro did hypothesize that some diseases were caused by the passage of "germs" from one thing to another. He described three processes for the transmission of contagion: direct contact, transmission via fomities (inanimate objects, such as clothing, which may pick up and preserve germs), and transmission from a distance. Fracastoro recognized that germs diseases occurring in different hosts and different processes of transmission for different germs.

The germ theory of disease had little immediate influence on medicine and human health. The belief was that cleanliness and hygeinic practices would control disease, and to a large degree sanitary practictise introduced in Europe after the Industrial Revolution lowered the incidence of disease. The great microbial epidemics were brought under control not by treatment with drugs but largely by sanitation and by the general rise in living standards. Sanitary measures were accompanied by a decrease of typhus morbidity and mortality. These sanitary measures were implemented by boards of health that did not believe in contagion, let alone in the germ theory of disease, Faith in the healing power of pure air, with much contempt for the germ theory of disease, was the basis of Florence Nightingale's reforms of hospital sanitation in the mid 1850s.

The last half of the nineteenth century marks a turning point in the epidemic history of the Western world. Transmissible diseases, were, of course, still plentiful; and scarlet fever, diphtheria, meningitis, and measles which had previously been masked to some extent by the more rapidly spreading and violent contagions— now attained greater prominence... But except for influenza, the pestilences which had, throughout preceding centuries, caused the most widespread destruction were distinctly declining and were becoming more limited in regional distribution.

The studies of Robert Koch changed the view of the cause of the disease and marked the beginning of modern medical microbiology. Koch, a German country physician, began his scientific studies isolated from any contact with scientific community, working alone with primitive tools and materials. As a result of his medical practice, Koch was well aware of the diseases of man and other animals. From 1873 to 1876, Koch conducted experiments to show that the spores of anthrax bacilli isolated from pure cultures could infect animals. Koch demonstrated for the first time that germs grown outside the body could cause disease, and that specific micro–organisms caused specific diseases. This was the beginning of Koch's illustrious career. He went on to determine the causative organisms for several other diseases, including tuberculosis and cholera.

Until late in the nineteenth century disease had been regarded as resulting from a lack of harmony between the sick person and his environment; as an upset of the proper balance between the yin and the yang, according to the Chinese, or among the four humors, according to Hippocrates. Louis Pasteur, Robert Koch, and their followers took a far simpler and more direct view of the problem. They showed by laboratory experiments that disease could be produced a will by the mere artifice of introducing a single specific factor— a virulent micro–organism into a healthy animal.

In describing the cause (etiology) of tuberculosis, Koch set forth the basic principles for establishing a cause and effect relationship between a given micro–organism and a specific disease.

Stated simply, Koch's four postulates for identifying the etiologic agent of a disease are:

1. The organism should be present in all animals suffering from the disease and absent from all healthy animals;
2. The organism must be grown in pure culture outside the diseased animal host;
3. When such a culture is inoculated into a healthy susceptible host the animal must develop the symptoms of the disease;
4. The organism must be reisolated from the experimentally infected animal and shown to be identical with the original isolate.

To obtain a complete proof of a causal relationship, rather than mere co-existence of a disease and a parasite, a complete sequence of proofs is necessary. This can only be accomplished by removing the parasites from the host, freeing them of all tissue elements to which a disease—including effect could be ascribed, and by introducing these isolated parasites into a healthy animal with the resulting reproduction of the disease with all its characteristic features. An example will clarify this type of approach. When one examines the blood of an animal that has died of anthrax, one consistently observes countless colourless, non-motile, rod like structures... When minute amounts of blood containing such rods were injected into normal animals these consistently died of anthrax, and their blood in turn contained rods. This demonstration did not prove that the injection of the rods transmitted the disease because all other elements of the blood were also injected. To prove that the bacilli, rather than other components of blood produce anthrax, the bacilli must be isolated from the blood and injected alone. This isolation can be achieved by serial cultivation... The serial transfers can be continued for 3 or as many as 50 passages and in this manner the other blood components can be eliminated with certainty. Such pure bacillins produce fatal anthrax soon after injection into a healthy animal, and the course of the disease is the same as if produced with fresh anthrax blood or as in naturally occurring anthrax. These facts proved that anthrax bacilli are the unique cause of the diseases.

Koch's postulates, which are applicable to plant as well as animal diseases, still form the basis for determining that a particular disease is caused by a given micro–organism. For example, the search for the cause of Legionnaire's disease in the 1970s followed Koch's 1890 postulates, resulting in the eventual identification of the bacterial etiologic agent. After many attempts, the bacterium *Legoinella pneumophila* was isolated from patients with this disease, grown in the laboratory, inoculated into test animals in which the organism caused the onset of disease symptomatology, and reisolated from the experimentally infected animals. Some modifications to Koch's postulates are required in cases when the disease is caused by opportunistic pathogens that are normally associated with healthy animals; when the experimental host is immune to the particular disease; when the disease process involves co-operation between multiple organisms, and when the causative agent cannot be grown in pure culture outside of host cells.

It is often difficult to find "volunteers" who are willing to be inoculated with presumed disease-causing micro–organisms. Times have changed since the early 1900s when Pettenkoffer in Germany and Metchnikoff in France, together with several of their associates, drank tumblerfuls of cultures isolated from fatal cases of cholera. Fortunately these investigators were resistant to cholera; although high numbers of cholera vibrios could be recovered from their stools and some of the self-infected experiments developed mild diarrhoea, none of these microbiologists developed true cholera. Today, medical reseachers try to avoid such experiments. Researchers have recently exposed large numbers of subjects to Rhinoviruses to study the transmission of the common cold but are not performing such experiments with more serious microbial pathogens. For example, although a virus has been isolated from individuals with AIDS (acquired immune deficiency syndrome) and grown in pure culture in the laboratory, it is not ethical to inject healthy individuals with the virus to establish a cause and effect relationship between the virus and AIDS. Despite these occasional difficulties, the philosophy of Koch's postulates for identifying the causes of infectious disease remains the fundamental basis of modern epidemiology.

Control of Infectious Disease

Chemotherapy

Responding to the growing awareness that micro–organisms are associated with disease processes, Joseph Lister, an English Quaker and physician, revolutionized surgical practice in 1867 by introducing antiseptic practices. The discovery in the early 1850s of anaesthesia and its administration to patients made surgery much easier but of course did nothing to reduce the incident of post-surgical disease that often was as high as 90 per cent, especially in military hospitals.

In the Nineteenth Century men lost their fear of God and acquired a fear of microbes.

Lister knew that in the 1840s Ignaz Semmelweis, a Hungarian physician who worked in maternity wards in Vienna, had shown that physicians who went from one patient to another without washing their hands were responsible for transmitting childbed fever, a disease that killed many women after childbirth. He was also aware that Pasteur had demonstrated that micro–organisms are present in the air. Lister used carbolic acid, phenol, as an antiseptic during surgery. He first used bandages soaked in carbolic acid to dress wounds from compound fractures in order to diminish the likelihood of infection. Later, he used a carbolic acid spray in addition to direct application of this compound during surgical procedures. Lister eventually discarded the practice of spraying after 17 years of trials as unnecessary, but he retained the use of direct application.

Various chemical formulations for preventing microbial growth and infection were described by Koch and his disciples, including Paul Ehrlich, who, like Pasteur, had been trained as a chemist. From 1880 to 1896, Ehrlich worked in Koch's laboratory and in 1896 he became director of the first of his own institutes, which he dedicated to finding "substances which have their origin in the chemist's retort", that is, substances produced by chemical synthesis to cure infectious diseases, Ehrlich's research between 1880 and 1910 established the early basis for modern chemotherapy. He established the correct formula for atoxyl, an arsenical, which was being considered for use in treating sleeping sickness, and developed almost a thousand new derivatives of this compound. Compound 606, salvarsan, proved to be effective in treating syphilis. The drugs developed by Ehrlich and his coworkers became known as magic bullets and were portrayed as being able to find and kill disease-causing germs.

A major breakthrough in chemotherapy occurred in 1929, when the Scottish bacteriologist Alexander Fleming, working in a London teaching hospital, reported on the anti-bacterial action of cultures of a *Penicillium* species. Fleming observed that the mold *Penicillium notatum* killed his cultures of the bacterium *Staphylococcus aureus* when the fungus accidentally contaminated the culture dishes, It is likely that contaminant of Fleming's cultures, which was to bring medical practice into the modern era of drug therapy, blew into his laboratory from the floor below, where an Irish mycologist was working with strains of *Penicillium*. Such a serendipitous event can change history, but in science it takes a special individual, like Fleming to recognize the significance of the observation. As Pesteur said, "Chance favours the prepared mind." After growing the fungus in a liquid medium and separating the fluid from the cells, Fleming discovered that the cell-free liquid was an inhibitor of many bacterial species.

After his studies on fowl pox, Pasteur directed his attention the study of anthrax. This irritated Koch, who considered anthrax to be his exclusive domain. There was an intense rivalry between Pasteur and Koch because of personal egotistical concerns and intense nationalistic pride. Because he enjoyed being the center of attention and controversy, Pasteur staged a very dramatic public demonstration to test the effectiveness of his anthrax vaccine. Witnesses to the demonstration were amazed to see that the 24 sheep, 1 goat, and 6 cows that had received the attenuated vaccine were in good health, but that all the animals that had not been vaccinated were dead of anthrax.

The art of the experimenter is to create models in which he can observe some properties and activities of a factor in which he happens to be interested. Koch and Pasteur wanted to show that micro–organisms could cause certain manifestations of disease. Their genius was to devise experimental situations that lent themselves to an unequivocal illustration of their hypothesis— situations in which it was sufficient to bring the host and the parasite together to reproduce the disease. By trial and error, they selected the species of animals, the does of infectious agent, and the route of inoculation, which permitted the infection to evolve without fall into progressive disease.

Pasteur, Koch, and their followers succeeded in minimizing in their tests the influence of factors that might have observed the activity of the infectious agents they wanted to study. This experimental approach has been extremely effective for the discovery of agents of disease and for the study of some of their properties. But it has led by necessity to the neglect, and indeed has often delayed the recognition, of the many other factors that play a part in the causation of disease under conditions prevailing in the natural world — for example, the physiological status of the infected individual and the impact of the environment in which he lives.

Even considering the impressiveness of this display, Pasteur's greatest success in developing vaccines was yet to come. In 1885 Pastur was able to announce to the French Academy of Science that he had developed a vaccine for preventing rabies. Although he did not understand the nature of the causative organism, Pasteur developed a vaccine that worked. Pasteur's motto was "seek the microbe", but the micro–organisms responsible for rabies is a virus, which could not be seen under the microscopes of the 1880s He, nevertheless, was able to weaken the rabies virus by drying the spinal cords of infected rabbits and allowing oxygen to penetrate the cords. Thirteen inoculations of successively more virulent pieces of rabbit spinal cord were injected over a period of two weeks during the summer of 1885 into Joseph Meister, a nine-year old boy who had been bitten by a rabid dog.

Since the death of the child was almost certain, I decided in spite of my deep concern to try on Joseph Meister the method which had served me so well with dogs... I decide to give a total of 13 inoculations in ten days. Fewer inoculations would have been sufficient, but one will understand that I was extremely cautious in this first case. Joseph Meister escaped not only the rabies that he might have received from his bites, but also the rabies which I inoculated into him.

This treatment met with a highly publicized, personal success for Pasteur. The development of the rabies vaccine capped Pasteur's distinguished career. Donations sent to Pasteur as a consequence of his discoveries were used to erect 1'-Institute Pasteur, the first priority of which was to provide the proper facilities for the production of vaccines. Today vaccination is used to protect children around the world against many diseases.

In a net-too-distant past children were expendable... It is now desired that all children survive. This ideal naturally makes it essential to discover ways that will substitute for the protective effects that childhood disease used to exert in the past. Vaccination to elicit without danger the immunity formerly resulting from the disease itself and sanitary procedures designed to eliminate infectious agents from the environment are the technological procedures through which man attempts to substitute for the natural mechanisms of biological adaptation.

Viruses

Only with advances in technique and improvement in apparatus is it possible to make fundamental advances through new ideas and observations. The development of bacteriologic filters and the discovery of viruses is a case in point.

Bacteriologic Filters

As an alternate to heat sterilization, unsuccessful efforts to remove bacteria from solutions by filtration through paper and similar materials led Chamberland and Pasteur to test and develop unglazed porcelain as the first successful bacterial filter (1871-1884). The Berkefeld filter of Kieselguhr (diatomaceous earth) was developed shortly thereafter 1891. Synthetic polymer filters of cellulose nitrate, cellulose acetate., polyester, and so forth, also introduced earlier, have come into common use only in the last two decades because of technical advances allowing quality control of pore size. It is of interest to not that these are essentially space-age products developed in part for the rapid removal of micro–organisms from jet and rocket fuels.

Discovery of Viruses

The tobacco mosaic disease agent was first discovered by Iwanowski in 1892 in bacteria-free filtrates of diseased plant juices. This finding, confirmed by Beijerinck in 1899, marked the beginning of studies on the so-called filterable agents.

The filterable agent causing foot and mouth disease in cattle, the first described animal virus, was discovered by Loffler and Frosch in 1898. The Yellow Fever virus of humans was discovered in 1900 by Walter Reed and his co-workers. Bacterial virusus, or bacteriophages, were discovered in 1915 by Twort in England and d' Herelle in France.

Viruses could not be grown in artificial media and Koch's criteria could not be specifically applied. Because these pathogens require a living host for propagation, rapid progress in their study developed only in recent years. Again, as in the Golden Age of Bacteriology, technology had to be developed, Outstanding were the development of the electron microscope, ultracentrifugation, tissue culture, and the application of sophisticated microchemical and biochemical techniques.

It is of interest to note that some filterable agents first thought to be viruses, such as *Mycoplasma, Rickettsia,* and *Chlamydia,* have been found instead to be bacteria. These organisms are fastidious in growth requirements: almost all require living host cells. However, ultra-structure and chemical analysis have shown these infectious agents to be bacteria.

Immunity

Ancient peoples immunized themselves against venomous snakes by introducing small amounts of venom into scratches in the skin. The Chinese used variolozation with dried material from dermic smallpox lesions for 20 centuries. This practice spread through Asia by trade routes and was well accepted in the Middle East. Later, apparently quite independently, Edward

Jenner (1749-1823) noticed that milk-maids who developed cowpox were immune to smallpox. He developed the concept of "vaccination" and was able to protect susceptible people by vaccinating them with cowpox. Pasteur developed a chicken cholera vaccine in 1877. He inoculated chickens with old attenuated cultures so that a mild disease rendered the chickens immune to virulent organisms. He called this vaccination after Jenner's procedure. Shortly aferwards, in 1881, applying the same concept, Pasteur prepared temperature attenuated anthrax grown at 42° to 43°C, and protected sheep by first injecting them with these bacteria before challenging them with virulent anthrax grown at lower temperatures. Salmon and Smith, in 1884-1886, used heat-killed cultures of hog cholera bacillus to develop resistance or immunity in swine against challenge by live virulent organisms. Pasteur developed rabies vaccine in 1886, again using the idea of injecting an attenuated living disease agent. In this case, Pasteur used dried animals spinal cords without, apparently, recognizing the viral form of the disease agent.

Two schools of thought arose in explanation for the increased resistance following vaccination. Metchnikoff developed, in the 1880s, the cellular theory of protection; Bordet and others proposed the humoral, or specific, antibody concept of immunity. There is now evidence that both theories are correct. The last two decades have resulted in the isolation and, in large measure, the structural description of the major humoral immune proteins, the immune gamma globulins. These are now commonly referred to as IgA, IgG, IgD, IgM, and IgE, The function of these various immunoglobulins are currently being intensively studied. For example, the co-relation of the well-known immediate allergic response with the release of histamine from mast cell as a result of the reaction between humorally injected antigen and mast cell bound IgE was elucidated in 1970. Much current work is also being devoted to the mechanisms of cellular interactions in immune reactions which occur only in infectious disease caused by bacteria, viruses, and fungi, but also in rejection reactions of tissue and organ transplants, and of cancer cells.

Antimetabolites

Many antimetabolites, which were pioneered in concept by Ehrlich in the mid to late 1800's, are now accepted household words, e.g., pencillin. The modern era of antibiotics developed only after Domagk reported in 1935 that Prontosil had a dramatic effect on streptococcal infections. It was soon discvered that prontosil was coverted in the body to sulfanillamide, the active chemical agent. The success of the sulfonamides, catalyzed new interest in chemotherapeutic agents. In the 1940's as the result of the stimulus of World War II, Florey and Chain and their associates reinvestigated Fleming's penicillin, isolated and characterized it and demonstrated its practical clinical value. As a result of millions of tests with thousands of organisms, we now have numerous other antibiotics active against almost all types of bacteria.

With the recognition that metabolic and structural differences, at the molecular level, exist between pathogenic micro–organisms and human or animal hosts, the rationale of developing new chemo-therapeutic compounds is now often based on exploiting these differences. There is every reason to believe that newer more specific and potent drugs will be discovered. However, chemotherapy has created new problems. Many previously susceptible organisms are now resistant to therapeutic levels of many widely used drugs. In addition, drug sensitization reactions or allergies occasionally develop, clinical syndromes are modified and the normal ecologic flora of the body is disturbed.

Impact of microbiology on genetics and bio-chemistry, and the development of molecular biology

The enormous advantages of homogeneous populations of cells for every conceivable type of investigation was soon realized. As a result, many of the epoch-making advances during the last century in cell physiology, bio-chemistry, and genetics have resulted from studies with micro–organisms or materials isolated from them. Over the last two decades, these advances have led to a precise way of investigating the structure and function of nucleic acids and proteins, which has become known as molecular biology. For example, the demonstration of the central role of DNA as the repository of genetic information resulted from the studies of Griffith in the 1920's that pneumococci could be transformed from one capsular type to the other, followed by the investigations of Avery and associates, who succeeded, during the 1940's in isolating DNA from the pneumococci and in demonstrating that it was actually the transforming factor. Final proof beyond doubt was provided by the demonstration of Hershey and Chase in 1952 that viral nucleic acid itself contained all the information necessary for virus multiplication. At the same time. Watson and Crick developed the double helix model of DNA, which led them to suggest that one of the complementary DNA strands could serve as a template for the systhesis of the other, thus providing a description of self-perpetuating gene replication and continuity.

Demonstration of the tramscription from DNA of infromation in the form of messenger RNA systhesized in complementary sequence to DNA soon followed, again in a microbial system. Messenger RNA was then found to be translated into polypeptides on ribosomes. By the mid 1960's Nirenberg and others had worked out the nature of the triplet RNA base sequences corresponding to the codon signals for each amino acid.

More recently, attention has focussed on the nature of the signals which specify initiation and termination of transcription and translation. Although the major portion of all of this research work was and is being carried out with microbial systems, evidence is constantly coming to hand which suggests that similar mechanisms operate in the cells of higher organisms, and more particularly, in mammalian cells. An important factor in this connection has been the development of the technique of tissue culture, which permits animal cells to be grown, cloned, and passaged like micro–organisms. Tissue culture of animal cells provided the essential experimental tool for the development of animal virology, which is now being studied so intensively at the molecular level that it should be possible to develop a rational system of anti-viral chemotherapy within the next one or two decades, thus enabling viral diseases to be brought under effective control just as bacterial diseases are already controlled by antibiotics. Finally, the application of the concepts concerning the regulation of gene expression which are being yielded by work with microbial systems to animal cells should, in the foreseeable future, provide insight into the fundamental control mechanisms operating in both normal and abnormal cell differentiation, including cancer.

The Microbial World

Microscopic forms of life can be found in vast numbers in nearly every environment known to man. They are found in the soil, in bodies of water, in the food and water we consume, and even in the air we breathe. Fortunately, the greater number of such micro–organisms are not harmful to man nor to the various forms of life in his world. The microbial world includes in its membership microscopic forms of life as well as those that are too small to be seen with an ordinary microscope.

Many micro–organisms occur as single cell, some are multicellular, and still others (such as viruses) do not have a true cellular appearance. Certain organisms, called *anaerobes*, are capable of carrying out their vital functions in the absence of free oxygen. However, the majority of organisms, referred to as aerobes, require free oxygen. Microbes are known which can manufacture the essential compounds for their physiological needs from atmospheric sources of nitrogen and carbon dioxide. Other micro–organisms, such as viruses and certain bacteria, are totally dependent for their existence on the cells of higher forms of life. The branch of science known as microbiology embraces all of these properties of micro–organisms and many more. For the most part, micro–organisms are unicellular and exhibit the characteristic features common to biological systems, such properties include:

Reproduction

The ability to multiply. Many micro–organisms are capable of both *asexual* and *sexual* forms of this basic process. (Sexual reproduction is simply defined as the union of nuclear material from two different cells, resulting in a new individual).

Metabolism

The ability to utilize a variety of substances as food and to obtain from them the needed energy for numerous cellular activities, the sum total of the chemical reactions associated with these activities. Metabolic reactions can be grouped into two categories, namely, *anabolism*, or constructive metabolism, and *catabolism*, or destructive metabolism. Reactions of the former type are associated with the synthesis of cellular components, needed for growth, reproduction, and repair, while those of the latter category include digestion and degradation of harmful substances.

Growth

Most micro–organisms, as well as other forms of life, generally increase in size due to materials produced intracellularly, a process of growth from within. This is in contrast to accretion, which is the accumulation of material from the external environment onto surfaces.

Irritability

The ability to respond to environmental stimuli, including temperature, acidity, intense light, and toxic substances.

Adaptability

Adjustment to environmental stimuli, Several microbial types survive unfavourable environments because they are capable of altering certain of their activities, for example the direction of their movement (motility), and because the secrete substances to detoxify harmful substances in their environments.

Mutation

The susceptibility to genetic change. Certain environmental factors, of either a normal or an experimental nature, can not only bring about changes in the genetic apparatus, but can increase the frequency of such reactions. Mutations, in general, appear with a definite degree of frequency.

Organization

It is clear from the processes performed by microbial forms of life that they must possess a certain level of organization and precision. Hence, it is quite appropriate to refer to microbes as small, organized units or *micro–organisms*.

Microbiology and its Subdivisions

Taxonomic Arrangement

Included among the subjects for study in the field of microbiology are algae, bacteria, fungi (molds and yeasts), protozoa, and viruses. As a matter of conveniences, the general branches of microbiology frequently are designated accordingly. This terminology, referred to as the taxonomic approach, includes:

Bacteriology

The study of bacteria,

Immunology

The study of a host's defence mechanisms (resistance) against disease, In general, this involves determining the contributions to resistance made by the specific activities of certain body cells, e.g., lymphocytes and phagocystes, and the various chemical components of body fluids such as antibodies, complement, and lysozyme. Modern immunology also is concerned with the diagnosis, prognosis, and prevention of disease.

Mycology

The study of fungi.

Phycology (Algology)

The study of algae.

Protozoology

The study of protozoa.

Integrative Arrangement

Another approach to the study of micro–organisms is the "integrative arrangement," as referred to by Luria and Darnell. Here the subdivisions of microbiology are directed toward analyzing the common or specialized characteristics or properties of microbes and their various interactions. Examples of these areas of study include:

Microbial Cytology

The study of microscopic, as well as sub-microscopic, details of microbial cells. The later type of investigation usually involves the teachniques of electron microscopy.

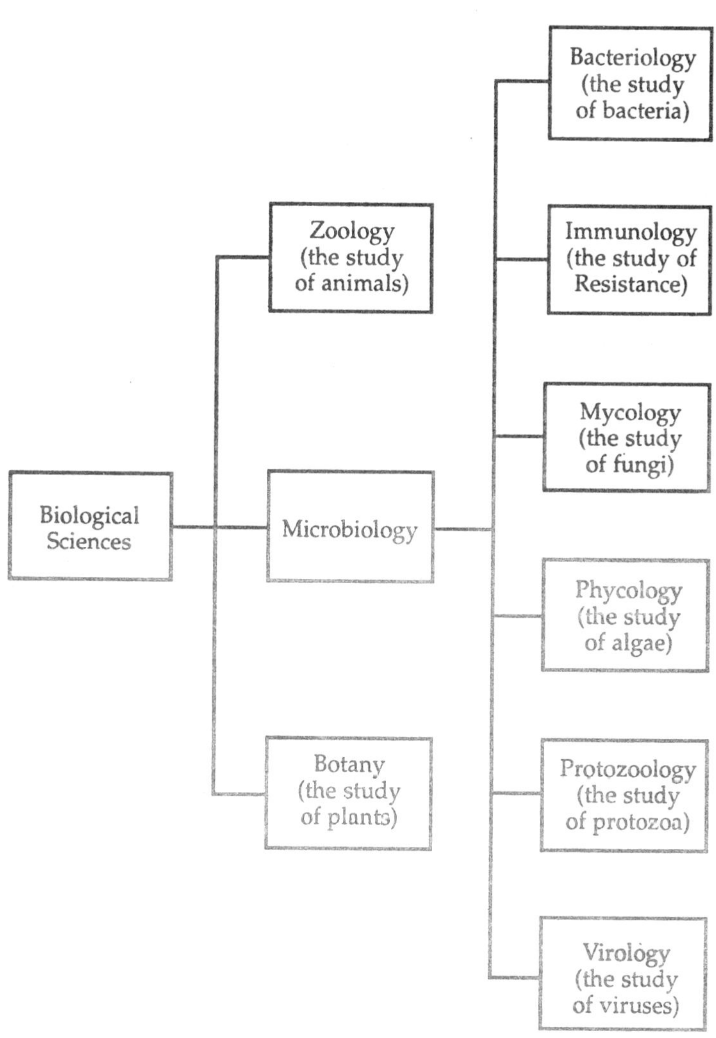

Fig. 1.5 Representation of the taxonomic approach to microbiology.

Microbial Ecology

The study of the relationships between micro–organisms and their environments. An investigation of the way micro–organism respond to unfavourable situations would be an example of this area of specialization.

Microbial Genetics

The study of inheritances. This area of investigation concerns itself with the activities of the nuclear elements of micro–organisms and how they regulate the growth and development of these forms of life; determine the effects of mutation-causing agents, i.e., mutagens; altering the genetic make-up of micro–organism; and uncovering the basis of antibiotic resistance in various microbes.

Microbial Physiology

The study of the functioning of micro–organisms. Metabolic activities, the effects of the environment on microbial synthetic pathways, and determining the nutritional requirements of different groups of microbes are a few of the types of investigations involving microbial physiology.

Microbial Taxonomy

The specialized area includes the naming and classifying of micro–organisms. Microbial taxonomy involves determining the similarities and differences among microbial species and using these data to formulate a classification system which shows the relationship of micro–organisms to one another.

Mention should also be made of two other areas of investigation, biochemistry and biophysics. Biochemistry is concerned with the chemical basis of living matter and the various reactions associated with them, while biophysics is devoted to the study of the principles of physics as they apply to all living matter.

These specialized fields are 'reductionist' sciences, analyzing basic processes in terms of electrons, atoms, and molecules. Many of these branches overlap. Moreover, one subdivision can include a number of others, each of which is concerned with a particular microbial group. For example, microbial genetics may be further divided and restricted to bacteria or to bacterial viruses, i.e., bacteriophages.

The education of a microbiologist today includes a background of general information in the majority of subdivisions. However, because of the tremendous accumulation of information in each specialization— which the individual cannot hope to master— the microbiologist must limit himself to one, or a select few, of the branches of microbiology. The branches he chooses to specialize in depend upon the type of microbial work he wishes to pursue. There are many different areas of applied microbiology.

Applied Areas of Microbiology

Many substances used by man in his everyday life are actually products of microbial activity. Microbe-dependent activities such as fermentation have been used since ancient times, but the improvements brought about by modern technology and the information obtained from the various areas of microbial specializations have made possible a closer control of micro–organisms and their activities in the field of industrial microbiology. This area of microbiology involves the efforts of chemists, engineers, and microbiologists toward efficient control of the conversion of raw materials into desirable end products by carefully selected micro–organisms.

The welfare of man is also affected by the particular activities of certain microbial species that interfere with the normal processes of the human body. Investigations concerned with such harmful effects of micro–organisms (pathogens) form the basis of medical microbiology. The following types of studies are included in this area of specialization: (1) determining the properties and disease capabilities of micro–organisms; (2) developing procedures to detect the presence of pathogens (diagnosis); (3) attempting to discover an antibiotic or antibiotics which will eliminate the disease agent or prevent it from exerting its harmful effects (antibiotic sensitivity testing); (4) developing vaccines against pathogens; (5) incorporating both chemical and physical methods for the managements and control of infectious diseases in the population as a whole. Today an ever- increasing number of heath team specialists are being trained in the principles of this specialized area.

Many animal parasites, such as hookworms and tapeworms, have microscopic stages in their life cycles, and courses concerned with the medical aspects of microbiology may include the study of these forms. Ordinarily, however, these metazoan (multicellular) *parasites* are considered separately in the specialized branch of *Parasitology*. In this textbook the parasitic protozoa, e.g. causative agents of diseases, including amebic dysentery and malaria, and the parasitic helminths will be considered in a specific division devoted to medical parasitology.

Microbiology and the Scientific Method

One goal of microbiology, as well as of every other branch of science, is to find explanations for observed phenomena and to show inter-relationships between them and related events. To achieve this aim, a type of organized common sense approach, referred to as the "scientific method," is used in one form or another. While all of the steps of this procedure may not be applicable to every aspect of microbiology, the essence of the method does direct the microbiologist to pose pertinent questions and to look for testable answers.

The scientific method involves making careful observations of a particular events and arranging them so that a generalization can be made to account for the observed phenomenon.

This particular step is called an *hypothesis*, or simply a "well-calculated guess." Once a hypothesis has been formulated, its validity must be tested. This phase of the scientific method is called *experimentation*. The simplicity or complexity of the hypothesis will determine the type and degree of experiments needed. Such experiments must be designed to test the pertinent point of the hypothesis, to include adequate controls for purposes of comparison, and to avoid the subjectivity or bias of the scientist himself. Experimental results must be repeatable by others. Experiments which cannot be repeated by other competent investigators are discarded.

Table— 1.1

The history of microbiology— an overview

Date	*Event*	*Historical Perspective*
1st Century B.C.	Varro suggests diseases due to invisible organisms	Caesar conquers Gaul.
1546	Fracastor's *De Contagione* makes first scientific statement of how infections are transmitted.	Henry VIII succeeds Edward VI.
1555	First use of word "physiology" in modern sense (Jean Femel).	
1590	First compound microscopes.	Shakespeare writes Henry VI.
1658	Kircher sees 'innumerable worms" under microscope.	
1660	Royal Society founded.	Rembrandt: "The Syndics of the Cloth Hall."
1665	*Philosophical Transactions* of the Royal Society first published.	
	Robert Hooke's *Micrographia*.	
	First drawing of cell (Hooke).	Delaware becomes a separate colony.
1676	Van Leeuwenheok discovers "little animalcules"; perfects lenses to magnify 300 times.	
1688	Redi publishes took on spontaneous generation of maggots from putrid flesh.	
1720	Bradley's germ theory.	London's Royal Academy of Music names George Frederick Handel as its director and presents Handel's oratorio *Ester*.

(Contd...)

Table 1.1 (Contd...)

Date	*Event*	*Historical Perspective*
1740	Buffon's orgar ic molecules" as infective agents floating in the air.	
1765	First drawing of cell division (Trembley).	
1765	Spallanzani shows that Buffon's "organic molecules" are distinct organisms.	
1770	Hill introduces new methods of staining and preserving specimens for microscopic study.	A spinning jenny that automates part of the textile industry is patented by English weavermechanic James Hargreaves. The Black Death strikes Russia and the Balkans in epidemic form.
1796	Janner inoculates James Phipps with cowpox.	Beethoven writes the "First Symphony."
1802	First use of word "biology" (Treveranus).	
1807	First achromatic microscope.	
1821	First international congress of biology (organized by Oken).	
1835	Bassi's theory of "living contagion" in silkworm disease.	
1838	Liebig establishes biochemistry.	The word "protein" is coined by Dutch chemist Gerard Johann Mulder, 36, who adapts a Greek word meaning "of the first importance." Gas ovens are installed at London's Reform Club. Coal or wood is the common cooking fuel in most of the world, but Arab nomads use camel chips, American Indians use buffalo chips, and Eskimos use blubber oil.

(Contd...)

Table 1.1 (Contd...)

Date	*Event*	*Historical Perspective*
1838-39	Schwann and Schleiden found modern cell theory: plants and animals composed of basically identical units.	
1844	Bassi asserts smallpox, bubonic plague, syphilis, spotted fever due to living parasites.	Potato crops fail throughout Europe, Britain, and Ireland as the fungus disease caused by *Phytophthora infestants* rots potatoes in the ground and also those in storage. Irish potatoes are even less resistant than potatoes elsewhere— up to half the crop is lost. Portland is founded in Oregon Territory near the junction of the Columbia and Willamette Rivers. The town is named after the 213-year-old city in Maine as two New Englanders let a flip of a coid decide in favour of Portland rather than Boston.
1845	Siebold recognizes protozoa as a single-celled organisms.	
1848	Semmelweiss demonstrates childbed fever is a form of septicemia and becomes a pioneer in aseptic technique.	
1850	Davaine asserts that anthrax is due to "bacterides" which he sees in the blood of dead sheep.	Dickens writes *David Copperfield*.
1854	Davaine sees 'monads" in stools of cholera patients.	
1857	Pasteur demonstrates that lactic acid fermentation is due to a living organism.	

(Contd...)

Table 1.1 (Contd...)

Date	*Event*	*Historical Perspective*
1858	Virchow's Doctrine "*Omnis cellula e cellula*" declares that "all cells come from cells."	Iowa State College is founded at Ames. Oregon State University is founded Corvallis. *Anatomy of the Human body, Descriptive and Surgical by* London Physician Henry Gray, is Published.
1860	First selective biological staining.	U. S. Civil War begins.
1864	Pasteur invents pasteurization (for wine).	
1867	Lister publishes work on antiseptic surgery.	
1869	Miescher discovers nucleic acid.	
		Washington D. C.'s Pennisilvania Avenue is paved with wooden blocks for a mile between 1st street and the Treasury Department building at 15th Street.
		"Ecology" is coined to mean environmental balance by German zoology professor Ernst Heinrich Haechel, 35, who is the first German advocate of Charles Darwin's organic evolution theory.
1876	Koch gives three-day demonstration of his work on anthrax, in which he had discovered the sequence of development.	
1877	Koch describes techniques of fixing, staining, and photographing bacteria. Also discovers *Bacillus anthracis* as cause of anthrax.	

(Contd...)

Table 1.1 (Contd...)

Date	*Event*	*Historical Perspective*
1879	Albert Neissen discovers *Neisseria gonorrhea* as cause of gonorrhea.	
1880	Laveran sees malarial parasite but is disbelieved.	
1880	Typhoid bacillus and leprosy agent discovered.	
1881	Koch words out method of culturing bacteria on gelatin, Pasteur, spurred by Koch's work, turns to study anthrax; publicly inoculates sheep at Melun with his "attenuated culture."	Dostoevsky writes brothers *Karamazov.*
1882	Koch discovers tubercle bacillus, and enunciates "Koch's Postulates."	
1882	Mechnikov launches phagocytic theory: "cellular theory of immunity."	
1883	First apochromatic microscopes.	
1885	Pasteur inoculates Joseph Meister for rabies. Theodor Escherich discover *E. coli.*	The first ready-to-use surgical dressings are introduced by Johnson and Johnson.
		Phagocytosis is discovered by Russian zoologist bacteriologist Ilya Ilich Mechnikov, 40.
1887	Buist, Edinburgh infirmary superintendent, sees pox virus and believes it is a form of bacteria.	
1888	Richet confers immunity on rabbits accidentally with serum from an infected dog.	
1897	Buchner discovers that cell-free yeast converts sugar to carbon dioxide and alcohol.	

(Contd...)

Table 1.1 (Contd...)

Date	*Event*	*Historical Perspective*
1897	Buchner demonstrates that cell-free yeast extract will catalyze glucose breakdown. Van Ermengem discovers *Clostridium botulinum.*	
1398	Beijerinck discovers and names tobacco mosaic virus; viral cause of foot-and-mouth disease demonstrated	
1898	Benda discovers and names the mitochondria, previously seen by Altmann.	First wireless communication between Europe and America.
1902	Richet discovers anaphylaxis.	
1902	Landsteiner investigates agglutination when blood from different human donors is mixed.	
1903	Sir Almroth Wright and others discover "opsonins" (i. e., antibodies) in blood of immunized animals.	
1905	Harden and Young show inorganic phosphate responsible for fermentative ability of yeast juice.	Helen keller is graduated magna cum laude from Radcliffe and begins to write about blindness.
1906	Schaudinn and Holffman discover *Treponema pallidum* as cause of syphilis.	
1906	Von Wassermann develops test for syphilis.	
1912	Ehrlich demonstrates first chemontherapeutic agent for a bacterial disease (syphilis).	World War I begins.

(Contd...)

Table 1.1 (Contd...)

Date	Event	Historical Perspective
1915	D' Herelle and Twort independently show existence of bacteriophage— viruses which destroy bacteria.	
1923	Landsteiner shows M and N factors in blood.	
1925	Keilin discovers cytochrome.	Al Capone takes over as boss of Chicago bootlegging from racketeer Johnny Torrio, who retires after sustaining gunshot wounds.
1926	Ultracentrifuge (Svedberg).	
1928	Elford demonstrates size of viruses (from 10 to 3000 micrometers).	The Great Depression begins.
1929	Lohmann identifies adenosine triphosphate (ATP) as necessary for the phosphorylation of sugar.	
	Fleming describes penicillin.	
1932	Sir Hans Krebs describes and names the citric acid cycle.	
1933	First electron microscope (Ruska).	Popular songs "Basin Street Blues" by Spencer Williams whose work was published in small orchestra parts 4 years ago; "Only a Paper Moon" by Harold Arlen, lyrics by E. Y. Harburg, Billy Rose; "Lazybones" by Hoagy Carmichael, lyrics by vocalist Johnny Mercer, 23; "Love is the Sweetest Thing; by Ray Noble; Stormy Weather-Keeps Rainin' All the Time" by Harold Arlen, lyrics by Ten Koehler.

(Contd...)

Table 1.1 (Contd...)

Date	*Event*	*Historical Perspective*
1935	Stanley crystallizes virus.	World War II begins.
1941	Beadle and Tatum establish "one gene-one enzyme" theory.	
1944	A very discovers "blueprint" function of DNA.	
1945	Role of mitochondria revealed.	Aerosol spray insecticides begin a revolution in packaging. The commercial "bug bombs" employ a Freon-12 propellant gas developed by two U. S. Department of Agriculture researchers in 1942. They have been used during the war to protect troops from malaria-carrying mosquitoes.
1948	Electrophoretic methods (Tiselius).	Korean conflict begins.
1952	Partition chromatography (Synge and Martin).	
1952	Hershey and Chase prove that DNA injected by bacteriophage is what disorganizes bacterium. Phage becomes the fruit fly of the molecular biologists.	The American Bandstand debuts in January on ABC network stations. Host Dick Clark, 22, will continue to emence the show for more
than		30 years.
1952		Films; Fred Zinneman's High Noon with Gary Cooper, Grace Kelly; John Houston's *The Red Badge of Courage* with Audie Murphy, Bill Mauldin.
1952	Sexual recombinaiton discovered in bacteria.	

(Contd....)

Table 1.1 (Contd...)

Date	*Event*	*Historical Perspective*
1952	Waksman discovers streptomycin, the first antibiotic effective against tuberculosis.	
1953	Lipmann discovers coenzyme A and its importance for intermediary metabolism.	
1953	Phase-contrast microscope.	
1953	Medawar shows tolerance to grafts can be conferred by inoculating newborn animal or embryo with antibodies from future donor.	
1953	Lederberg and Zinder discover transduction.	
1953	Crick and Watson propose a structure for deoxyribonucleic acid— a double spiral.	
1954	Enders, Weller, and Robbins discover poliomylitis viruses in cultures of various types of tissue.	
1957	Virus structure determined. Interferon discovered.	Sputnik, the first earth satellite, launched by Russia.
1958	Lederberg makes discoveries concerning genetic recombination and the organization of the genetic material of bacteria.	Cocoa puffs breakfast food, introduced by General Mills, is 43 per cent sugar.
		Transatlantic jet service is inaugurated by Pan American World Airways and British Over-seas Airways (BOAC).
1959	Komberg and Ochoa awarded Nobel Prize for discovery of enzymes that produce artificial DNA and RNA.	

(Contd...)

Table 1.1 (Contd...)

Date	*Event*	*Historical Perspective*
1961	Nirendberg, using artifical DNA, synthesizes a protein molecule.	
1962	Role of thymus in immunity established.	
1962	Crick, Watson, and Wilkins make discoveries concerning the molecular structure of nucleic acid and its significance for information transfer in living material.	President Kennedy killed.
1964	Bloch and Lynen work out the mechanism and regulation of the cholesterol and fatty acid metabolism.	U. S. Marines land in Vietnam.
1966	Rous discovers tumor- inducing viruses.	
1966	Huggins identifies hormonal treatment of prostatic cancer.	
1968	Holley, Khorana, and Nirenberg win a Nobel Prize for interpreting the genetic code and its function in protein synthesis.	U. S. Natural gas consumption begins to exceed new gas discoveries and reserves for interstate pipelines begin falling.
		U. S. First class postal rates climb to 6 cents; up from 5 cents per ounce in 1963.
1969	Delbruck, Hershey, and Luria win a Nobel Prize for the replication mechanism and genetic structure of viruses.	First man on the moon.
1971	T. O. Diemer identifies viroids.	
1972	Edelman and Porter: Nobel prize for work concerning the chemical structure of antibodies.	U. S. Apollo 16 astronauts Charles M. Duke, Thomas K. Mattingly, and John W. Young blast off April 16 from Cape Kennedy.

(Contd...)

Table 1.1 (Contd...)

Date	Event	Historical Perspective
		A human skull found in northern Kenya by Richard Leakey and Glynn Isaac allegedly dates the first humans to 2.5 million B. C.
1974	Claude, De Duve, and Palade receive a Nobel Prize for discoveries about the structural and functional organization of the cell.	
1975	Baltimore, Dulbecco, and Temin awarded a Nobel Prize for researching the interaction between tumor viruses and the genetic material of the cell.	
1976	Gajdusek and Blumberg did research leading to Nobel Prize for a test to show hepatitis viruses in donated blood and to an experimental vaccine against the disease.	U. S. A Bicentennial. Carter becomes President.
1978	Arber, Smith, and Nathans given Nobel Prize for discovery of restriction enzymes and their application to the problems of molecular genetics. Austrial wins award for first vaccine against pneumococcal pneumonia.	
1979	Henle identified first virus regularly associated with human cancer.	
1980	First U. S. patent issued for the process of producing biologically functional molecular chimeras; inventors: Stanley N. Cohen and Herbert W. Boyer.	
1980	Nobel Prize for Chemistry; Paul Berg, Walter Gilbert, and Frederic Sanger for development of a rapid way to determine the chemical makeup of DNA.	
1982	Epstein and Barr receive award for showing relationship between EBV and Burkitt's lymphoma.	World's Fair in Knoxville, Tennessee.

The next stage of the scientific method is the *theory*. This is an explanatory hypothesis that has been supported by various types of observations and experiments. A good theory can be used to predict new facts, to show relationships between phenomena, and to relate new information as it is uncovered. Quite often the term *law* is used interchangeably with *theory*. A distinction must be made, however, for a law is a theory which has attained universal acceptance. Many theories do not achieve this distinction.

Present Approaches

Some approach to the presentation of a subject, however, must be followed, and the following plan appears to the author to be most logical. Some knowledge of *methodology* involved in microbiology is essential. Methodology as such may be related to microbiology in that knowledge of techniques should aid the student in acquiring and assimilating knowledge of all phases of microbiology. Students should become familiar with micro-organism types and the general relation of micro-organisms to each other. *Methods of naming and classifying* micro-organisms as well as criteria employed in classification processes should be studied. Naming and classifying micro-organisms should deal with general structures and also with some modes and life processes.

The *nutrition* of micro-organisms is probably as good a place as any to begin a real survey study, although here artificial conditions are generally imposed. General nature and action of enzymes appear to be a logical follow-up of nutrition. Enzyme studies should be followed by concepts of metabolism, which should be approached from standpoints of compound degradation obtaining energy and of biosynthesis. It seems that *growth* and *reproduction* should follow biosynthesis. These processes relate primarily to individual organisms.

The *ecology of micro-organisms* should be placed after other considerations. Processes as they apply to individual organisms or particular groups should be studied first. The knowledge obtained from studies of individual groups should be related to microbial populations on a broader scale. In other words, any knowledge of micro–organisms should be related to other micro–organisms in an ecological evaluation. Both in vitro and in vivo ecological involvement should be considered, and it appears most logical to consider in vitro ecology first. Ecological conditions will include the roles and of micro–organisms in geology, oceans, soils, foods, and water contamination, sewage disposal and industrial uses of microbial metabolic products. Microbial ecology in relation to geological, marine, and soil conditions was working for numerous millennia prior to its involvement in food, sewage, and water supplies as related to the human race. Micro–organisms were probably involved in water throughout their cycle but were possible pathological contaminants of water supplies only very recently.

The utilization and control of micro–organisms is very recent and is extremely useful to man, to other animals, and to plants. Control measures that have been utilized, however, affect only a small part of nature's supply of micro–organisms and alter natures's overall ecology only slightly. Methods of control and utilization that have proved beneficial to mankind, however, should be presented.

Microbial ecology in relation to plants and animals is relatively new. All nature has been able to live together, although at many times members of one or more species have eliminated other species. At present, microbes and animals manifest a unique ecology in that many microbial species live on the skin, in mouths, in nasal tracts, and in alimentary tracts of animals with no apparent detriment to the animal. On the other hand, micro–organisms of many types have become parasitic and pathogenic to animals, and animals, in turn, have built up resistance against invasion and infection by them. Certain micro–organisms have also become adapted to forms of plant parasitic life, but the plant has not been able to produce specific immune substances to counterattack them. Discussion of parasitism of micro–organisms for plants and animals, the reaction of infected plants and animals to invasion by organisms and the specific mechanisms by which micro–organisms infect should come near the end of a study of biology of micro–organisms.

A set of principles should aid the student in acquiring an overall evaluation of microbiology

As a general introduction to micro–organisms, certain principles be presented. These should be kept in mind as the student studies the biology of micro–organisms. Some general principles are outlined here:

1. Although wide variations in structure are evident in micro–organisms, the only two clear-cut divisions are between viruses and other micro–organisms and between the procaryotic and eucaryotic cell types of non-virus forms. A division in the viruses exists between deoxyribonucleic acid (DNA) and ribonucleic acid (RNA) forms.
2. Although small and usually termed unicellular, each micro–organism is capable of carrying out all necessary life processes independently of other cells.
3. All micro–organism structures are built on carbon skeletons; therefore, an adequate carbon source is essential. Micro–organisms have ulilized essentially all organic carbon compounds and CO_2 as carbon sources.
4. All life, including microbial, must have energy in order to carry on life processes; micro–organisms have exploited numerous methods by which essential energy may be obtained. In most methods of obtaining energy, external electron donors and external electron acceptors are essential.
5. Energy must be obtained by processes carried out *inside* the cell. Principal energy sources for micro–organisms are substrate phosphorylation, photosynthetic phosphorylation, and oxidative phosphorylation. In all cases, high energy phosphate compounds are involved.
6. Micro–organisms accomplish much (or possibly all) of what they do by means of enzyme action.
7. Micro–organisms must possess a method of replication in order to continue existence.

8. Micro–organisms probably existed on earth for long periods of time in the absence of other forms of life. They maintained a balance in nature while alone on earth and continue to do so in the presence of other living forms.
9. Micro–organisms have drastically altered their environments through the ages and, in turn, have been altered by their environment.
10. Micro–organisms utilize metabolic products and degraded structures of all forms of life, including those of other micro–organisms, as nutrients.
11. Micro–organisms are ubiquitous, and through out history of life on earth they have become adjusted to many types of life including parasitism on plants, animals, and each other.
12. A peculiar type of ecology exists between micro–organism and animals in that when micro–organisms have parasitized animals, the animals have developed an active defense mechanism against the organism and its metabolic products.

The preceding principles are meant only as guidelines for the student, and the student should not spend too much time in attempting to convert them to verbal memory. Instead, the student is encouraged, along with the instructor, to refer to them often and gradually build up a general pattern of thought concerning the broad picture presented by the biology of micro–organisms. These principles and the summary metabolic chart will be referred to frequently throughout the text. It is my desire that at the end of the course in which this text is used, the students will be able to put the parts of the microbial life puzzle together in an orderly manner. Constant building up to the pattern will greatly aid students in visualizing the picture. An important step in understanding the value of and subject is understanding what constitutes the subject. What is meant by a micro–organism?

Small and large are relative terms

The term *micro* means small and in general may be used to describe any small object. The term macro, on the other hand, means large and is employed in characterizing large objects, the term *bios* is used in connection with *living* material as contrasted with non-living or lifeless material. From these concepts the term *microbiology* would connote biology of small things.or, stated differently, the study of small living forms. In its general application, then *microbiology may be defined as the study of small forms of life.* The next question is the way in which small and large are defined. For example, a fly is small when compared to an elephant but large when compared to a protozoan or a bacterial cell. A protozoan cell, on the other hand, may be considered quite large when compared to a molecule. The bacterial cell is usually much smaller than the cell of a protozoan, but it is still many times larger than the largest molecule. The virus particle, on the other hand, is generally much smaller than a bacterial cell but still larger than a molecule. Although the molecule is small in terms of a bacterial or virus cell, it is relatively large in comparison to atoms, which in turn, are composed of still smaller entities. The terms *micro* and *macro,* however, have some meaning when delineated and used in reference to a living forms. In general, although not universally, the micro-organism is thought of as the one-celled organism. Unicellular forms have all life processes within one cell, whereas multicellular forms possess a differentiation of function among different cells. The biology of micro–organisms may therefore be considered as a study of the biology of one-celled organisms. This line of demarcation will not be complete because there are some forms represented in the study of microbiology that may not be single celled. The study of algae and fungi may not be the study of single cell, but each cell is capable of life in its own individual and characteristic manner. A study of microbiology would be incomplete without an inclusion of the study of viruses. Viruses, however, may not be thought of as cells in the general sense of the word but as entities all their own. The study of microbiology, then, cannot be limited to the study of one-celled plants and animals but must be conceived of as a system that incorporates viruses and possibly other forms of life that are not considered unicellular plants or animals.

The term *bios,* or the concept of life, is in reality undefinable. The biologist, whether microbiologist, or microbiologist, would not define life as such. An informed and thoughtful biologist would probably describe a number of life processes that differentiate the living from the non-living rather than attempt to define life itself. *Growth, reproduction, irritability, movement*, and other processes have been construed as peculiar to living organisms. These may be considered life processes and, along with others, have often been designated as such. In a different method of designation, however, one may think of photosynthesis, nitrogen fixation, oxidation and reduction, and the inter-conversion of large molecules as being characteristic of life. These processes must be carried on in an orderly fashion and at controlled temperatures in living forms. Differentiation between elements in living and non-living forms is considered in connection with the nutrition of micro–organisms. The student, then, should develop an appreciation for differences between living and lifeless, although it may not be possible to write out a definition or a clear concepts of fundamental differences.

Significance of Studies

There are many reasons for studying microbiology. Until recently, considering the span of history, the study of micro–organisms was not possible. The actual study of microbiology had to await the discovery of a methodology that will be discussed in the following chapter. Discovery of many phases in the development of microbiology followed the development of techniques, as will be seen in late chapters. With our present knowledge, however, there are many reasons for studying microbiology and for utilizing it as a basic biological science. In the first place a micro–organism is a complete living entity within itself, although it contains only one cell. This cell, in so-far as possible, is independent of existence or functions of other cells. The biologist is able to study micro–organisms in their simple and complete forms in single cells. This is not to say that one cell can usually be isolated and studied alone, but it does mean that one can study a large number of like cells and obtain an overall picture of structure and function of *single cell*. The filed of microbiology is broad. For example, there are many different cell types and probably some living forms of entities in microbiology that are not cells at all. As will be seen later, the nutrition of micro–organisms encompasses that found in all forms of life. Micro–organisms carry on all essentially all types of metabolism and compound changes that occur elsewhere in life and have some that are peculiar to micro–organisms themselves.

There is a tendency on the part of educators to outline pigeon-hole knowledge. This tendency is especially inviting in the study of microbiology because of the nature of the study. Historically, microbiology is a very young science and has been described primarily by hit-and-miss methods. Characteristically, most micro-biologists are specialists in particular areas of study, and this emphasis has been derived from their training in most cases. As in order sciences, no one person can be acquainted with all fields and ramifications of microbiology. Some courses in microbiology emphasize one aspect of the science rather than an overall picture. If one wishes to examine the contents of microbiology, however, it will be necessary first to try to obtain a panoramic view. This can be done only by the study of micro–organisms without undue emphasis on their importance as disease-causing agents, their relation to industry, or and other characteristic role of practical importance they may play. The overall picture of micro–organisms will go far in acquainting the student with a panoramic view of biology in general. One important factor to keep in mind is that the study of micro–organisms will probably circumvent every form of metabolic life and go far beyond the others. We are in continuous contact with micro–organisms in every walk of life in which we are engaged.

As a further emphasis on the importance of a knowledge of micro–organisms, the following should be considered. When we wash our hands, comb our hair, tie our shoes, dress, or undress, we are contacting micro–organisms. When we walk on the soil of the good earth, we trample on literally billions of them. They are prevalent in the air we breathe and in great numbers in the food we eat. They are distributed over the surface of the human body as well as throughout the alimentary tract. They live in the water and milk we drink and on the gum we chew. Fortunately for the human race, however, only a small percentage of micro–

organism are harmful. The majority, as well shall see, are beneficial, and the human race could not long exist without them. Specific areas of livelihood will be outlined with relation of micro–organism to those areas, but this does not mean that micro–organisms and their components can be divorced from other forms of life. Some important areas of microbiology will be outlined here, and relationships of organisms to some way of life of mankind will be described in later chapters.

Cell structures of both plants and animals have been extensively studied. In higher plants and animals there are differentiations that are fairly evident, but relationships between structures of plants and animal cells are seen best in micro–organisms. Studies of the cell structure of micro–organisms have indicated clearly that there is a continuous gradation between plant and animal cells and that no fine line of demarcation can be established, as was formerly thought. As will be emphasized later, bacterial cells, along with the cells of blue green algae, are entirely different from the cells of other living organisms. Structures of cells and relations to living and non-living materials are best seen through studies of micro–organisms.

A knowledge of basic principles of *physiology* and *metabolism* had to await developments of technology in micro–organisms studies. Total organisms have been studied extensively, but a study of the physiology of a particular cell was an essentially impossible task. Development in the field of microbiology brought with it an opportunity to study a complete organism within one cell. It is usually thought that many life processes which go on within bacterial or other micro–organism cells are very similar to or even identical with those which are carried on in cells of higher plants or animals. Although general requirements of both plants and animals were known for many years, the real *differentiation between nutritional requirements of individuals was* more clearly defined in studies of micro–organisms. The nature of nutrients that organisms attacked and of metabolites that they gave out was about all that was known concerning nutrition and metabolism for many years.

Micro–organisms play a tremendously important role in the *fertilization and defertilization of the soil.* Nitrogen is taken out of the air by micro–organisms and put into the soil in a form that is available to plants. This may be done independently, or it may be done in connection with growth on the roots of leguminous plants. An understanding of *methods of inheritance* has advanced tremendously since micro–organisms have been used for tools in genetic studies. Bacteria have only single strands of chromatin material and are thus more easily studied genetically than some other micro–organisms. They also possess the advantages of rapid growth, multiplication, and inexpensive maintenance. In addition, bacteria and easily kept and manipulated. A study of virus genetics has added significantly to our knowledge of heredity, even though some do not consider viruses among micro–organisms.

Until recently it was very difficult to determine whether or not a certain food product contained materials that manufacturers claimed were present in it. Assay tests required the raising of animals for long periods of time and tremendous expense. Micro–organisms, however, can be used for *assay agents* very easily, and assay tests are run now within days or even within hours. Assays carried out by micro–organisms are inexpensive, accurate, and reproducible. It is tremendously important to have a knowledge of bacteriology or microbiology in general in the preparation of biologicals. Not only do workers need to know how to prevent the growth of micro–organisms as contaminants but they also need to know how some biologicals are produced. Many are produced from the growth of micro–organisms themselves. A knowledge of chemotherapy and immunology is uniquely significant in this respect. The necessity for a knowledge of microbiology in *medicine* need not be stressed at this point. Although relationships between micro–organisms and infections were not known until recent years, events the layman is now aware of microbial infections. In the field of medical technology, technologists need to have a thorough knowledge of micro-organisms in order to perform their duties. Nurses need to be aware of micro–organisms and their characteristics in order to prevent contamination by them, keep materials clean, and sterilize instruments for use in surgery.

A knowledge of microbiology is tremendously important *in the household*; knowledge of this nature is more practicable than theoretical and technical in most cases. Housewives are acquainted with the fact that refrigeration will prevent food spoilage. On the other hand, they are probably not acquainted with peculiarities of psychrophilic organisms but know, nevertheless, that refrigeration will preserve foods for certain periods of time. It is evident, therefore, that a knowledge of microbiology is important in essentially every vocation in life. We may even extrapolate microbiology to include nursery rhymes. Let us consider Little Miss Muffett who sat on a tuffet eating her curds and whey as an example. Althought her being frightened by the spider had no direct relationship to microbiology, preparation of curds and whey did. Curds and whey probably resulted from the action of lactic acid bacteria in milk. Consider also the butcher, the baker, and the candlestick maker. As the nursery rhyme states, they all fell out of a rotten potato. Micro-organisms were instrumental in causing rot in the potato, and the butcher and backer were also in professions in which a knowledge of micro–organisms's actions would definitely be beneficial. The butcher needed to know how to preserve and take care of his meat, and the baker was aware of the action of yeast on rising bread and also of spoilage of foods that had been prepared. The candlestick maker may not be primarily involved, but there are certain micro–organisms that will destroy materials that are used in candles. Another illustration of microbiology in story book lore in the case of Jack and Jill who *went up the hill* to get a pail of water. Natural drainage of the terrain would let waste products run away from dwellings and not contaminate springs with pathogenic enteric micro–organisms.

We can see, therefore, that microbiology is a broad subject and micro–organisms are connected with all forms of life. No vicissitude of life is isolated from micro–organisms. Micro–organisms are important to the merchant as well as to the farmer and rancher. Foods may spoil, and molds may attack shoes and other leather goods. To the oil industry, micro–organisms play a significant role in hydrocarbon (petroleum) breakdown.

Furthermore, from its beginning the shipping industry has been plagued by microbial action. In addition to eroding concrete structures and wooden parts of ships, they furnish rough surfaces on the hulls of ships to which barnacles can attach. There are many branches of industry in which micro–organisms play a role. Some knowledge of them is important in related business professions, although again a knowledge of microbiology may be practicable rather than theoretical. Micro-organisms in relation to spoilage of foods is well known. A few other examples of micro–organism of metabolism in relation to industry inlcudes the clogging of pipes in sugar manufacture, eroding of concrete in construction, production of animal food, interference in performance of aircraft fuel, and preservation of paper in the printing industry. Disposition of sewage and other wastes is highly influenced by microbial growth.

A large number of reasons for studying micro–organisms have been presented. Certain limitations of information from microbiological studies, however, should also be noted. There is little or no differentiation of tissues in micro–organisms, for example, since each cell represents a living entity all its own. In higher plants and animals there are divisions and differentiations among cells. Although there are numerous different micro–organisms types and each organism has its own particular cell type, there is usually no differentiation of function among cells, and when differentiation is made, it is usually unnecessary for growth and propagation of the organisms. There are also many cell types found elsewhere in nature that are not present in micro–organisms. Although one can study the physiology of an individual cell or group of like cells in micro–organisms, it is difficult to study physiological processes that result from a combination of a group of cells that form tissues or organs. Never impulses, blood circulation, muscular contraction, and many other physiological processes are not represented in the field of micro–organisms. Processes similar to nerve impulses and muscle contraction may be realized in chemotaxis and flagellar motion, but differences between these processes and nerve and muscle responses in animals are apparent. *Parasitism and ecology reach their acme in microbial life.* This peak, however, is reached only in the relation of one complete entity to another, and there is usually no differentiation of cell function. Co-operation, or synergism, is a well known phenomenon in microbiology, but this represents complete organisms in relation to one another rather than tissues in relation to other similar or different types. In short, microbiological studies do not account for the functioning of *parts of an organism in relationship to the whole.*

Although the field of genetics has been greatly advanced by studies of micro–organisms, there are certain limitations here also. Processes of mitiotic and miotic division cannot be studied in procaryotic micro–organisms because if they are apparent, they are well camouflaged. In the case of procaryotic cells of bacteria and blue-green algae, chromatin material is found only in one strand, and duplication of chromosomes is not a recognized phenomenon. Although sexuality has been recognized in phases of certain micro–organisms, this phenomenon is not well-known and is less well-developed than it is in higher plants and animals. When sexuality is present in unicellular forms, it appears to be a function of the whole cell instead of a differentiated part.

One essential condition for studying life habits of micro–organisms is found in a medium in which organisms will grow. Although organisms may grow profusely in their natural habitats, this is usually not the best place for studying them. This situation is especially true in regard to bacterial cultures. In the case of virus studies it is not possible to obtain much information concerning them in their natural habitat with present technological methods. If one is studying protozoa, algae, some fungi, and a few other micro–organisms, it may be possible to observe them under natural conditions of growth. The study of bacteriology, however, has generally been pursued under conditions of growth in artificially prepared. One method of studying micro–organisms is by culturing them and determining conditions best suited for their growth. For many years microbiologist studied micro–organisms without obtaining them in pure cultures, and they often observed mixtures rather than pure forms. Contamination was usually found in experiments performed by Koch, Pasteur, and other early bacteriologists. It was difficult, and for years considered an impossible task, to isolate a particular type of organism and separate it from the others present.

In cases of studies relating to higher plants and animals, individuals can usually be studied as they appear in nature, or the natural environment can be closely duplicated. The study of flora and fauna of natural habitats is the rule, but it is truly the exception in microbiology.

2

Sterilization

Sterilization may be defined as the complete destruction of all living organisms in, or their removal from, materials by means of heat, filtration, or other physical or chemical methods.

Plugged test tubes, flasks, bottles, etc., and Petri dishes must be sterilized before use to destroy all living organisms adhering to the inner surfaces. Pipettes are placed in containers and heated to sterilize both inner and outer surfaces. Likewise, all culture media must be sterilized previous to use in order to destroy all contaminating organisms present. Studies on single bacterial species or pure cultures could not be made if the glassware and culture media were contaminated with other organisms previous to use. When once sterilized, glassware may be kept in a sterile condition indefinitely if protected from outside contamination. The same applies to culture media, if in addition to sterility, evaporation can be prevented.

The usual methods employed for the sterilization of laboratory materials involves the use of heat. Three types of heat sterilizers are used in bacteriology for the destruction of living micro–organisms: (1) the hot-air sterilizer, (2) the Arnold sterilizer, and (3) the autoclave.

Hot-air Sterilizer : This is a dry-air type of sterilizer (**Fig. 2.1**). It is constructed with three walls and two air spaces. The outer walls are covered with thick asbestos to reduce the

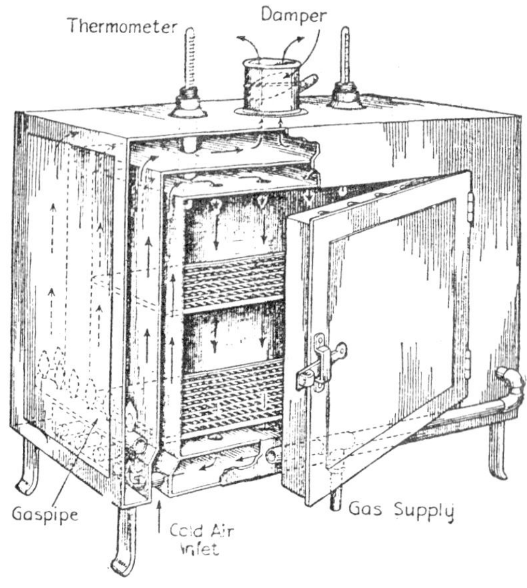

Fig. 2.1 Hot-air sterilizer.

radiation of heat. A burner manifold runs along both sides and rear between the outside and the intermediate walls. Convection currents travel a complete circuit through the wall space and interior of the oven, and the products of combustion escape through an opening in the top.

The hot-air sterilizer is operated at a temperature of 160 to 180°C. (320 to 356°F.) for a period of 1½ hr. If the temperature goes above 180°C., there will be danger of the cotton stoppers charring. Therefore, the thermometer must be watched closely at first until the sterilizer is regulated to the desired temperature. The necessity of watching the sterilizer may be avoided by having the oven equipped with a temperature regulator.

The hot-air sterilizer is used for sterilizing all kinds of laboratory glassware, such as test tubes, pipettes, Petri dishes, and flasks. In addition, it may be used to sterilize other laboratory materials and equipment that are not burned by the high temperature of the sterilizer. Under no conditions should the hot-air sterilizer be used to sterilize culture media, as the liquids would boil to dryness.

Arnold Sterilizer : It is well known that moist heat is more effective as a sterilizing agent than dry heat. This is believed to be due to the following reasons: (1) moist heat has greater penetrating power; and (2) death of organisms is believed to be caused by a coagulation of the proteins of the protoplasm. An increase in the water content of the protoplasm causes the proteins to coagulate at a lower temperature.

The Arnold makes use of streaming steam as the sterilizing agent (**See Fig. 2.2**). The sterilizer is built with a quick-steaming base that is automatically supplied with water from an open reservoir. The water passes from the open reservoir, through small apertures, into the steaming base, to which the heat is applied. Since the base contains only a thin layer of water, steam is produced very rapidly. The steam rises through a funnel in the center of the apparatus and passes into the sterilizing chamber.

Sterilization is effected by employing streaming steam at a temperature of approximately 100°C. (212°F.) for a period of 20 min. or longer on three consecutive days. The length of the

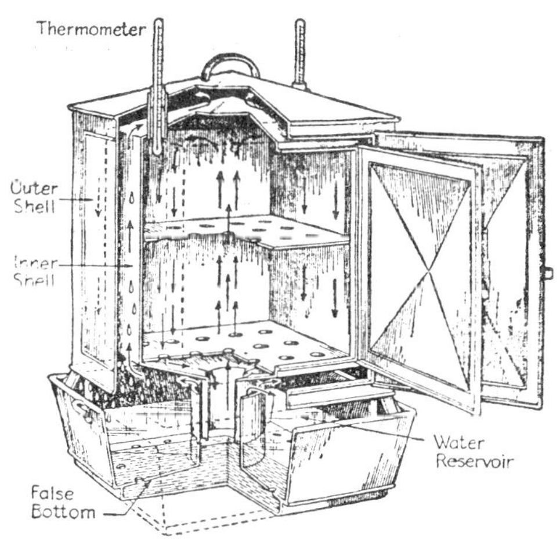

Fig. 2.2 Arnold sterilizer.

heating period will depend upon the nature of the materials to be treated and the size of the container. Agar, for example, must be first completely melted before recording the beginning of the heating period.

It must be remembered that a temperature of 100°C. for 20 min. is not sufficient to destroy spores. A much higher temperature is required to effect a complete sterilization in one operation over a relatively short exposure period.

The principle underlying this method is that the first heating period kills all the vegetative cells present. After a lapse of 24 hr. in a favourable medium and at a warm temperature, the spores, if present, will germinate into vegetative cells. The second heating will again destroy all vegetative cells. It sometimes happens that all spores do not pass into vegetative forms before the second heating period. Therefore, an additional 24-hr. period is allowed to elapse to make sure that all spores have germinated into vegetative cells.

It may be seen that unless the spores germinate the method will fail to sterilize. Failure may be due to the following causes: (1) The medium may be unsuited for the germination of the spores. Distilled water, for example, is not a favourable environment for the growth of bacteria. Therefore, it will not permit spores to germinate into vegetative cells. (2) Spores of anaerobic bacteria may be present, which will not germinate in a medium in contact with atmospheric oxygen.

The Arnold is used principally for the sterilization of gelatin, milk, and carbohydrate media. Higher temperatures or longer single exposures in the Arnold may hydrolyze or decompose carbohydrates and prevent gelatin from solidifying. Obviously, such media would then be unsatisfactory for use.

Autoclave : The autoclave is a cylindrical metal vessel having double walls around all parts except the font **(Fig. 2.3)**. It is built to withstand a steam pressure of at least 30 lb. per sq. in.

The principle of the method is that water boils at about 100°C., Depending upon the vapour pressure of the atmosphere. If the atmospheric pressure is increased, the temperature will be

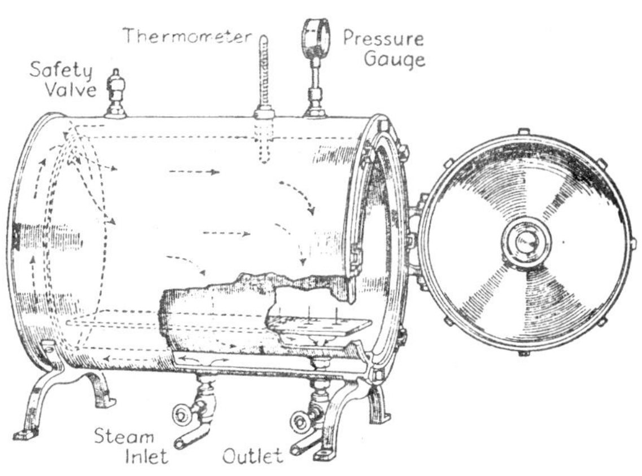

Fig. 2.3 Autoclave sterilizer.

increased. Therefore, if the steam pressure inside the closed vessel is increased to 15 lb. per sq. in. (2 atm.), the temperature will rise to 121.6°C. The relationship between pressure and temperature is shown in Table 2.1.

The autoclave is operated usually at 15 Ib. steam pressure for a period of 15 min., which corresponds to a tempertaure of 121.6°C. This temperature is sufficient to destroy both vegetative cells and spores in one operation.

Table 2.1

Pressure, lb. per sq. in.	*Corresponding tempertaure* °C.	°F.
5	107.7	227
10	115.5	240
15	121.6	250
20	126.6	260
25	130.5	267
30	134.4	274

Certain precautions must be observed to prevent sterilization failures. The most important single cause is incomplete evacuation of air from the chamber. Observation of the pressure

(table Contd...)

gauge alone is not sufficent. The proper degree of temperature must also be taken into consideration. The temperature figures given in Table 2.1 are true only if all air is evacuated from the sterilizing chamber.

The temperature of a mixture of steam and air at a given pressure is less than that of pure steam alone. This means that, even though the autoclave is kept at the desired pressure, the temperature may not be high enough to give complete sterilization. The actual temperatures attained in the autoclave under ordinary conditions of proper and improper usage, according to Underwood (1934), are given in Table 2.2.

Table 2.2 Temperature with Various Degrees of Air Discharge

Gauge, pressure, lb.	*Pure steam, complete air discharge*		*Two-thirds air discharge 20-in. vaccum*		*One-half air discharge, 15-in, vaccum*		*One-third air discharge, 10-in. vaccum*		*No air discharge*	
	°C.	°F.	°C.	°F.	°C.	°F.	°C.	°F.	°C.	°F.
5	109	228	100	212	94	202	90	193	72	162
10	115	240	109	228	105	220	100	212	90	193
15	121	250	115	240	112	234	109	228	100	212
20	126	259	121	250	118	245	115	240	109	228
25	130	267	126	259	124	254	121	250	115	240
30	135	275	130	267	128	263	126	259	121	250

Another important precaution to be observed is that the steam must have access to the materials to be sterilized. If the steam is prevented from penetrating the materials, the method will be of doubtful value. For example, suppose that it is desired to sterilize some cotton contained in a bottle. If the bottle is closed with a rubber stopper, the steam cannot reach the cotton. The process will be no more effective than a hot-air sterilizer kept at 121.6°C. for a period of 15 min. It has already been seen that such a temperature and time interval is insufficient to destroy spores in a dry-air sterilizer. On the other hand, if the mouth of the bottle is covered with one or two thickness of muslin, permitting the steam to penetrate, then the cotton will be sterilized.

The autoclave is used to sterilize most types of solid and liquid media with and without carbohydrate, gelatin media, distilled water, normal saline solution, discarded cultures, contaminated media, aprons, rubber tubing and gloves, etc. This is the type of sterilizer employed commercially for processing canned foods. For more information see Hoyt, Chaney, and Cavel (1938) and Under wood (1937).

Sterilization By Filtration

Some solutions cannot be sterilized by heat without being greatly altered in their physical and chemical properties. Serum in culture media is easily coagulated by heat. If the serum content is high enough, the medium becomes changed from a liquid to a solid preparation. Certain physiological salt solutions containing the unstable compound sodium bicarbonate are ruined if heated. The bicarbonate easily loses carbon dioxide and is converted into the more alkaline sodium carbonate. Enzymes and bacterial toxins in solution are easily destroyed by heat. These are but a few examples of many that might be mentioned.

Preparations containing heat-sensitive compounds are best sterilized by the process of filtration. The types of filters employed for this purpose include Chamberland and Jenkins porcelain filters, Berkefeld and Mandler diatomaceous earth filters fritted-glass filters, asbestos filters, and collodion membranes or ultrafilters.

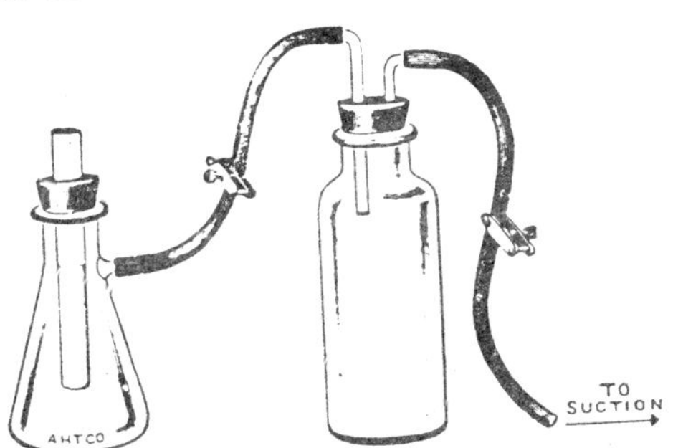

Fig. 2.4 Chamberland-Pasteur filter assembled in a filter flask and ready for filtration.

Porcelain or Chamberland Filters : Porcelain filters are hollow, unglazed cylinders, closed at one end. They are composed of hydrous aluminium silicate or kaolin with the addition of quartz sand and are heated to a temperature sufficiently low to avoid sintering. These filters are prepared in graduated degrees of porosity, from LI to LI3. Cylinders having the largest pores are marked LI; those having the smallest pores are designated LI3. The finer the pores, the slower will be the rate of filtration. The LI and L2 cylinders are preliminary filters intended for the removal of coarse particles and large bacteria. The L3 filter is probably satisfactory for all types of bacterial filtration.

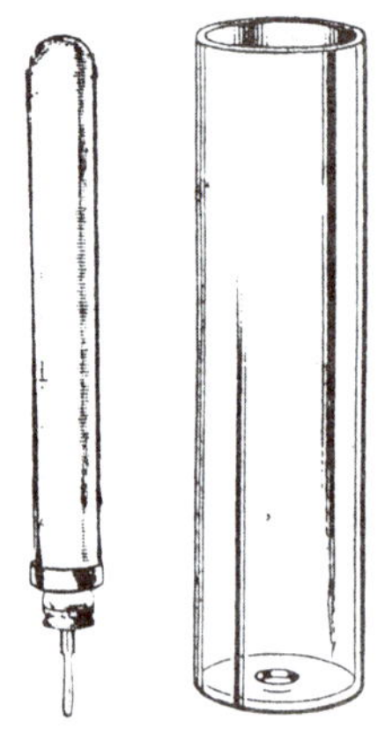

Fig. 2.5 Left, Berkefeld filter. Right, glass mantle for use with either Berkefeld or Mandler filter.

Berkefeld Filters : Kieselguhr is a deposit of fine, usually white siliceous powder composed chiefly or wholly of the remains of diatoms. It is also *called diatomaceous* earth and infusorial earth.

Berkefeld filters are manufactured in Germany. They are prepared by mixing carefully purified diatomaceous earth with asbestos and organic matter, pressing into cylinder form, and drying. The dried cylinders are heated in an oven to a temperature of about 2000°C. to bind the materials together. The burned cylinders are then machined into the desired shapes and sizes.

The cylinders are graded as W (dense), N (normal), and V (coarse), depending upon the sizes of the pores. The grading depends upon the rate of flow of pure filtered water under a certain constant pressure.

Mandler Filters : These filters are similar to the Berkefeld type but are manufactured in this country. They are composed of 60 to 80 per cent diatomaceous earth, 10 to 30 per cent asbestos, and 10 to 15 per cent plaster of Paris. The proportions vary, depending upon the sizes of the pores desired. The ingredients are mixed with water, subjected to high pressure, and then baked in ovens to a temperature of 980 to 1650°C. to bind the materials together.

The finished cylinders are tested by connecting a tube to the nipple of the filter, submerging in water, and passing compressed air to the inside. A gauge records the pressure when air bubbles first appear on the outside of the cylinder in the water. Each cylinder is marked with the air pressure obtained in actual test.

A convenient arrangement of apparatus for filtering liquids through a Mandler or Berkefeld filter is shown in Fig. 2.6. The reduced pressure is indicated by the manometer. The liquid to be filtered is poured into the mantle, and the filtrate is collected in a graduated vessel, from which it may be withdrawn aseptically. Filtration may be interrupted at any time by stopping the vacuum pump and opening the stopcock on the trap bottle to equalize the pressure.

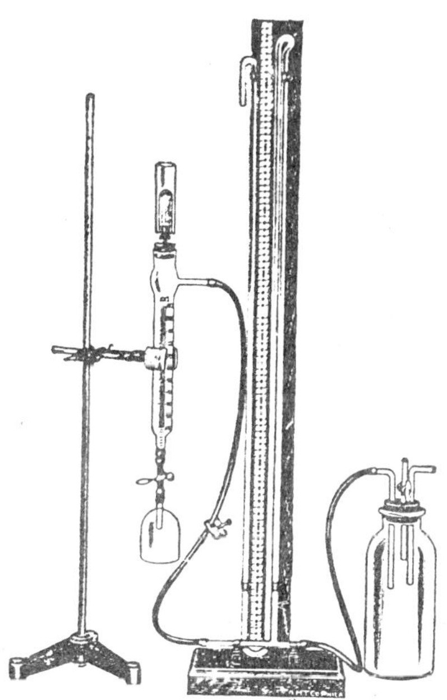

Fig. 2.6 A convenient arrangement for assembling a Berkefeld or Mandler filter for filtration.

Fritted- glass Filters : Filters of this type are prepared by fritting finely pulverized glass into disk form in a suitable mold. The pulverized glass is heated to a temperature just high enough to cause he particles to become a coherent solid mass, without thoroughly melting, and leaving the disk porous. The disk is

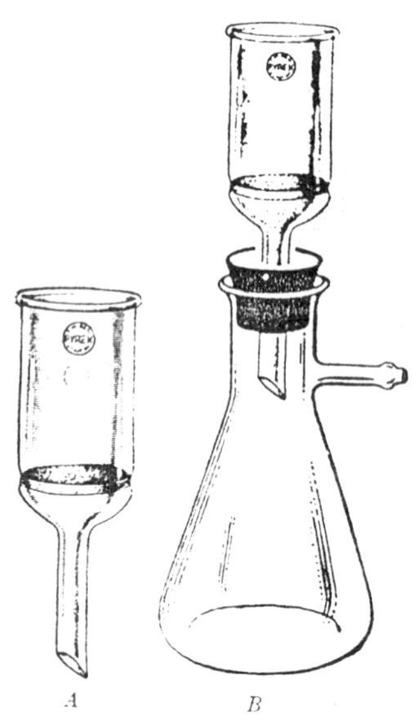

Fig. 2.7 A, fritted-glass filter. B, filter assembled in a filter flask and ready for filtration.

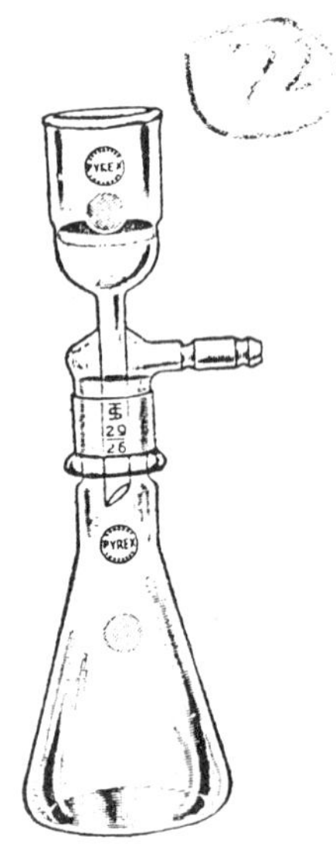

Fig. 2.8 Fritted-glass filter coupled to a flask through a ground-glass joint.

then carefully fused into a glass funnel and the whole assembled into a filter flask by means of a rubber stopper (**Fig. 2.7**). Another arrangements is the coupling of the filter to the flask through a ground-glass joint (**Fig. 2.8**) thus eliminating the use of a rubber stopper.

The filters are marketed in five degrees of porosity as follows: EC (extra coarse), C (coarse), M (medium), F (fine), and UF (ultrafine).

Bacteriological filters are generally employed under conditions of reduced pressure, Bush (1946) recommended filtration through glass filters by the use of positive pressure. Positive pressure not only reduces or eliminates evaporation of the filtrate, but greatly facilitates the interchange of receivers particularly important in bacteriological filtrartions which must be handled aseptically.

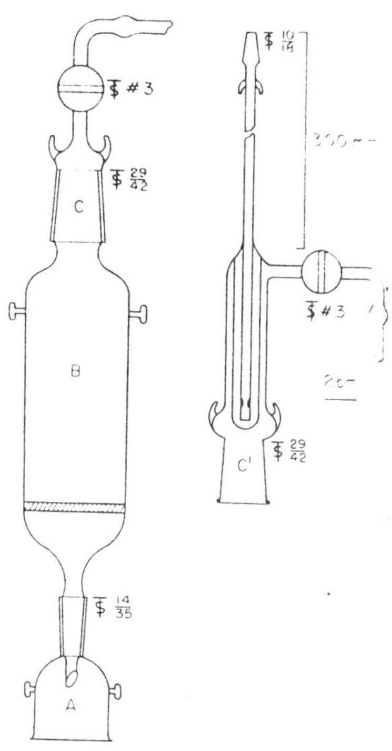

Fig. 2.9: Arrangement of bacteria filter for positive-pressure filtration. (*After Bush.*)

A convenient arrangement is shown in Fig. 2.9. The main body of the filter *B* contains a fritted-glass disk. A shield *A* protects the receiver from dust, and a pressure head carries a stopcock, An alternate pressure permits the retention of pressure head *C′* contains a built-in mercury manometer. The stopcock on *C* (C′) permits the retention of pressure after the apparatus is detached from the source of compressed air. An ordinary rubber pressure bulb is satisfactory for producing pressures up to at least 450 mm. Mercury. If the ground- glass joints are well lubricated and the parts held together with strong rubber bands or springs, the apparatus should hold this pressure for days.

Asbestos Filters : The best-known filter employing asbestos as the filtering medium is the Seitz filter (**Fig. 2.10**). The asbestos is pressed together into thin disks and tightly clamped between two smooth metal rims by means of three screw clamps. The liquid to be filtered is poured into the metal apparatus, in which the asbestos disk is clamped, and the solution drawn through by vaccum. the filtering disks are capable of effectively retaining bacteria and other particulate matter. At the end of the operation, the asbestos disk is removed, a new one inserted, and the assembled filter sterilized. This feature makes the Seitz filter very convenient to use, since no preliminary cleaning is necessary.

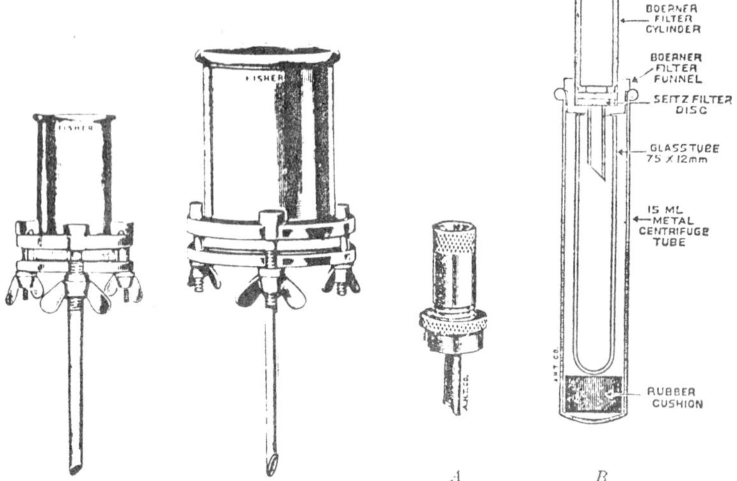

Fig. 2.10 Seitz filters with asbestos disks in place.

Fig. 2.11 A, Boerner centrifugal filter; *B*, Boerner filter assembled in a 15-ml. metal centrifuge tube, with glass collecting tube inside.

A modification of the Seitz filter, utilizing centrifugal force instead of suction or pressure, has been suggested by Boerner (**Fig. 2.11**). The filter consists of a cylinder and a funnel-shaped part with stem, which holds the filter pad supported on a wire gauze disk. The cylinder screws into the funnel with the filter disk pressed between them. The assembled filter fits closely into the top of a 15-ml. metal centrifuge tube, with the knurled collar of the funnel portion resting on the top of the metal tube. The filtrate is collected in a glass tube inside the cup. The filter can also be used for vacuum filtration in the conventional manner by inserting the stem through a rubber stopper fitted to a filter flask.

Jenkins Filter : This filter consists of a metal mantle holding a soft rubber sleeve and a porcelain filter block. The porcelain block is held in the rubber sleeve and made watertight by screwing together two metal parts. The filter block is not fragile. It is washed after each use, dried, and inserted in the mantle. The mantle is fitted with a rubber stopper, wrapped in paper, and sterilized in an autoclave. For use, the filter is attached to a flask (**Fig. 2.12**).

The filter is designed to be used for the sterilization of small quantities of liquids.

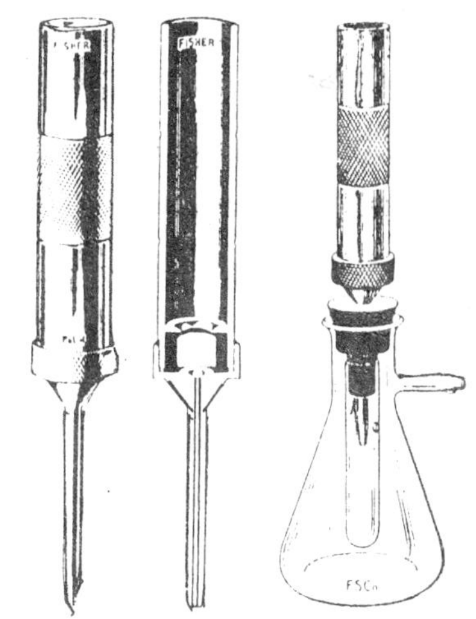

Fig. 2.12 Jenkins filter. Left, completely assembled; center, cross section of filter; right, filter assembled and attached to a filter flasks by a rubber stopper. Small tube inside of flask is used to collect the filtrate.

Ultrafilter : Ultrafiltration generally means the separation of colloidal particles from their solvents and from crystalloids by means of jelly filters known a ultrafilter.

The early jelly filters were composed of gelatin and of silicic acid, but these have been replaced by collodion in membrane or sac form, or collodion deposited in a porous supporting structure. The supporting structure may be filter paper in sheet and thimble form, unglazed porcelain dishes and crucibles, Büchner funnels, filter cylinders, etc.

Collodion : Several types of collodion are employed for ultrafiltration. The earliest type was prepared by dissolving pyroxylin or soluble guncotton in a mixture of 1 part of alcohol and 3 parts of ether. A more popular type is prepared by dissolving pyroxylin in glacial acetic acid. Pore size may be controlled by increasing or decreasing the pyroxylin content or by adding various liquids such as ethylene glycol and glycerol to alcohol-ether collodion. Elford (1931) recommended a new type of collodion for the preparation of a graded series of filters which he termed *Gradocol* membranes. The filters are prepared by incorporating a definite amount of amyl alcohol with an alcohol-ether collodion and then adding graded amounts of water or acetic acid to increase or decrease the permeability of the filters. Since membranes prepared by this procedure are quite strong, it is not necessary to deposit the collodion in a porous supporting structure.

Sacs may be prepared by during the collodion into a test tube or beaker, inverting and twirling continuously so that the excess drips out, until a thin, even coating is formed. The tube or beaker is then plunged into cold water to jell the collodion. After most of the solvent has been washed away, the sac can be loosened from the glass mold and removed. Collodion membranes in sheet form may be prepared by cutting the sacs at right angles at the closed end and then lengthwise.

Membranes employing filter paper as the porous supporting structure are usually prepared with acetic collodion. Pore size of the membranes depends upon the strength of the collodion. A strong collodion gives finer pores than a weak collodion. It is a simple matter to prepare a graded series of filters. The paper foundation gives the membranes a strength that, under certain conditions, will withstand a pressure of 20 atm. Or more.

Collodion filters are extensively used for the isolation of ultramicroscopic viruses and for determination of their diameters.

Electrical Charge of Filters. The filtration of solution of enzymes, toxins, immune bodies, viruses, etc., usually results in a loss of some of the active material. If the active material is present in very low concentration, it may be completely removed from solution.

Filters composed of porcelain diatomaceous earth, fritted glass, and asbestos consist chiefly of metal silicates and carry negative electrical charges.

The metal (Mg+ +, Al + + +, Ca+ +, etc.) cations or positive ions are more soluble than the silicate anions or negative ions and show a greater tendency to pass into solution. When a liquid is filtered, positively charged particles will react with the negative silicate ions and negatively charged particles will react with the positive metal ions. Since the metal charged particles will react with the positive metal ions. Since the metal ions are soluble, they will react with the negatively charged particles and pass through the pores the filter into the filtrate. On the other hand, the insoluble silicate ions will react with positively charged particles and remain fixed to the walls of the filter pores.

Adsorption of compounds from solution can be very effectively demonstrated by means of basic and acid dyes, such as toluidine blue and picric acid, respectively. Toluidine blue is a thiazine dye having the same chromophore as methylene blue and thionine. It ionizes as follows:

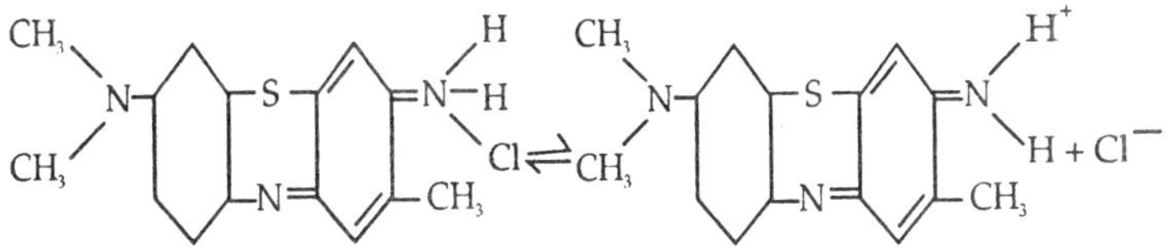

The cation is blue, and the chloride anion is colourless.. When a solution of this dye is filtered through one of the silicate filters, the blue cations will react with the negative silicate ions and remain fixed to the pores of the filter. The chloride anions will combine with the metal cations of the silicates to form soluble metal chlorides and pass into the filtrate. If more than sufficient dye is present in the solution to react with all the silicate ions in the filter pores, the excess will pass through, imparting a blue colour to the filtrate. If, on the other hand, the amount of toluidine blue is insuffícent to take care of all the silicate ions, the dye will be completely removed from the solution and the filtrate will be colourless. The reaction is reversible, however, since the passage of distilled water through the filter saturated with dye results in a blue colour in the filtrate.

If a solution of an acid dye is used instead of the basic toluidine blue, the results will be quite different. The dye will not be adsorbed by the filter material but will pass through the pores into the filtrate.

Picric acid or trinitrophenol is an acid dye in which the nitro group is the chromophore. It ionizes as follows:

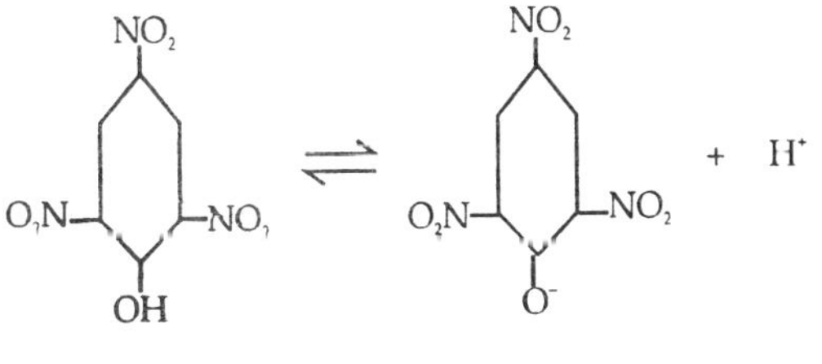

In this case, the cation is colourless and the anion is yellow. When a solution of this dye is filtered one of the silicate filters, not a trace of it will be adsorbed because an exchange of ions results in the formation of soluble picrates, which pass through in the filtrate.

Michaelis (1925) found that collodion filters were non-ionogenic but that they also carried a negative charge. Their negative charge was believed to be due to the adsorption of negative ions. Elford (1933) found that proteins in solutions, adjusted to different pH values by hydrochloric acid and sodium hydroxide, were most strongly adsorbed in the isoelectric zone. On the other hand, proteins in solutions buffered with M/ 15 phosphate, instead of adjusting the pH with hydrochloric acid and sodium hydroxide, are absorbed on the acid side of the isoelectric zone. The negatively charged collodion now preferentially adsorbs the positively charged proteins. Elford concluded that the effect is probably associated with some specific influence of the phosphate ion.

Amphoteric Nature of Proteins and Amino Acids : An important characteristic of proteins and amino acids is that they contain both acidic (COOH) and basic (N/NO_2) groups. In acid solutions, the compounds act as bases; in basic solutions, they act as acids, Representing the formula of an amino acid as R. $CHNH_2COOH$, the reactions with acids and bases are as follows:

With an acid,

$$\begin{array}{c} R \\ | \\ H—C—NH_2 \\ | \\ COOH \end{array} + HCl \rightarrow \begin{array}{c} R \\ | \\ H—C—NH_3Cl \\ | \\ COOH \end{array}$$

On ionization, this gives

$$\begin{array}{c} R \\ | \\ H—C—NH_3Cl \\ | \\ COOH \end{array} \rightleftharpoons \begin{array}{c} R \\ | \\ H—C—NH_3^+ \\ | \\ COOH \end{array} + Cl^-$$

The acid reacts with the basic amino group. The amino acid molecule has a positive charge and, therefore, behaves as a base.

With a base,

$$H-\overset{\displaystyle R}{\underset{\displaystyle COOH}{C}}-NH_2 + NaOH \rightarrow H-\overset{\displaystyle R}{\underset{\displaystyle COONa}{C}}-NH_2 + H_2O$$

On ionization, this gives

$$H-\overset{\displaystyle R}{\underset{\displaystyle COONa}{C}}-NH_2 \rightleftharpoons H-\overset{\displaystyle R}{\underset{\displaystyle COO^-}{C}}-NH_2 + Na^+$$

The base reacts with the acid carboxyl group. The amino acid molecule has a negative electrical charge and, therefore, behaves as an acid. Compounds of this nature that are capable of reacting with both acids and bases are said to be amphoteric.

Isoelectric Point : According to the classical theory, amphoteric compounds are supposed to dissociate into ions on either side of a pH point known as the isoelectric point. The isoelectric point has been defined as that point where the ionization of the amphoteric compound is at a minimum, expressed in pH. Opposed to this concept is the more recently developed view known as the *zwitterion* hypothesis, which states that the isoelectric point is that point where ionization is at a maximum. The difference in the two theories is indicated in the following formulas for isoelectric alanine:

$$H-\overset{\displaystyle CH_3}{\underset{\displaystyle COOH}{C}}-NH_2 \qquad H-\overset{\displaystyle CH_3}{\underset{\displaystyle COO^-}{C}}-NH_3^+$$

I. Classical II. Zwitterion

Formula I represents a molecule that is not dissociated as either an acid or a base. The neutrality of the molecule is assumed to be due to the absence of dissociation. Formula II is also neutral, but the neutrality is assumed to be due to complete ionization of the acidic and basic groups. Regardless of which theory is correct, results obtained on the addition of an acid or a base are the same in both cases.

	Addition of HCl	Isoelectric	Addition of NaOH
Classical	$H{-}\overset{CH_3}{\underset{COOH}{C}}{-}NH_3 + Cl^- \leftarrow$	$H{-}\overset{CH_3}{\underset{COOH}{C}}{-}NH_2 \rightarrow$	$H{-}\overset{CH_3}{\underset{COO^-Na^+}{C}}{-}NH_2$
Zwitterion	$H{-}\overset{CH_3}{\underset{COOH}{C}}{-}NH_3 + Cl^- \leftarrow$	$H{-}\overset{CH_3}{\underset{COO^-}{C}}{-}NH_3^+ \rightarrow$	$H{-}\overset{CH_3}{\underset{COO^-Na^+}{C}}{-}NH_2$

The isoelectric point of a protein is not necessarily the neutral point (pH 7.0). As a matter of fact, most proteins which have been studied have isoelectric points on the acid side of neutrality. The isoelectric points of a few of the common protein are as follows:

	Isoelectric point (pH)
Casein (milk protein)	4.7
Egg albumin	4.6
Gelatin	4.7
Hemoglobin	6.8
Serum albumin	4.8
Serum globulin	5.6

A knowledge of the isoelectric points is of considerable value in the filtration of solutions containing proteins, amino acids, bacterial toxins, enzymes, viruses, antitoxins, etc. If a solution is acid with respect to its isoelectric point, the active constituent will behave as a base and possess a positive electrical charge. The filtration of such a solution through a silicate filter, which has a negative charge, will result in the complete or partial absorption of the active constituent on the walls of the filter pores. To avoid this, it would be necessary to use a filter having a positive charge, or to change the reaction of the active constituent to the alkaline side of its isoelectric point.

The adjustment of a solution to correspond to the acidic or basic side of the isoelectric point can be carried out only provided the change in pH of will not result in a destruction of the active material. After filtration, the pH of the filtrate should be readjusted to correspond to the optimum pH range of the active component. If the active material is very sensitive to slight changes in pH, a filter having an appropriate electrical charge should be selected instead.

Cleaning Filters : Some filters are discarded after each use and new ones employed; others are intended to be cleaned after each filtration and, with proper care, may be used repeatedly. Collodion membranes are easily prepared, and the Seitz asbestos disks are relatively low in cost. These filters are intended to be used once, then discarded. On the other hand, porcelain, diatomaceous earth, and fritted-glass filters are too expensive to be used only once, but are easily cleaned.

Porcelain filters are cleaned by placing them in a muffle furnace and raising the temperatures to a red heat. This burns the organic matter in the pores and restores the filters to their original condition.

Filters of the Berkefeld and Mandler types are cleaned by placing the cylinders in a special metal holder connected to a faucet. The flow of water is reversed by passing through the cylinder from within outward. This should be continued until all foreign matter has been washed away from the filter pores. Albuminous or similar materials remaining in the pores of the filters are likely to be coagulated by heat during the process of sterilization, with the result that the filters will be clogged. Filters in this condition are useless for further work.

Clogged filters may be cleaned in various ways but probably most conveniently by continuous suction of full-strength Clorox, or similar solution, for 5 to 15 min. This treatment quickly dissolves the coagulated material and restores the usefulness of the filter. Thorough washing is necessary to remove the last traces of the oxidizing solution.

Fritted-glass filters may be cleaned by treatment with concentrated sulfuric acid containing sodium nitrate. The strong acid quickly oxidizes and dissolves the organic matter. Thorough washing is necessary to remove the last traces of acid.

Sterilization of Filters : With the exception of collodion membranes, the various filters are assembled in their appropriate holders, wrapped in paper, and autoclaves. Dry heat cannot be used because of the destruction of rubber fittings for the filter flasks.

Since acetic collodion solutions are sterile, it is not necessary to sterilize such membranes in which filter paper is used as the porous supporting structure, if aseptic precautions are observed in their preparation.

3

Culture Methods

Although the microscope furnished the first means for studying bacteria, it did not permit the bacteriologist to describe micro-organisms except on the basis of morphology. Under natural conditions bacteria grow as communities that may include dozens of species, many of which are morphologically inseparable. In order to study individual species, it is necessary to cultivate them without other types of organisms being present. Bacteria in such an environment are said to be in a "pure culture." The isolation of a bacterial species is dependent upon the ability of the bacteriologist to supply an environment suitable for growth which is free from all other forms of life

The Theory of Spontaneous Generation : Early attempts to provide a sterile environment failed because the workers were not aware of the ability of bacterial spores to survive prolonged boiling or of the presence of bacteria and mold spores in airborne dust. Actually, the methods that led to the successful culturing of bacteria arose from the controversy over the spontaneous generation of life. This ancient theory that life could arise without antecedents from non-living materials had been thoroughly discredited as far as larger animals and insects were concerned. The rapid appearance of molds, yeasts, and bacteria in freshly prepared juices and recently cooked foods, however, helped to substantiate the theory when it was applied to micro-organisms. The idea of life arising spontaneously had a certain appeal because it helped to explain inadequate observations and poorly performed experiments. For example, between 1745 and 1750 John Needham, an English priest, found that micro-organisms promptly developed in extracts of meat and grain which he had heated in stoppered bottles. Such evidence was difficult to refute because it could be argued that a failure to obtain similar results was due to inadequate conditions. Thus, when an Italian abbot, Lazzaro Spallanzani, believed that he had successfully demonstrated that Needham's results were due to insufficient heating and imperfect seals, he was in turn criticized because he had excluded air which would be necessary for the development of living creatures. For nearly a century the controversy centered around the problem of admitting air to a sterile solution without also letting in bacteria and molds. Passage of air through solutions of acid and alkali or heated tubes was challenged on the basis that a "vital force" was thereby removed from the air. In 1854, Schroeder and Von Dusch demonstrated that air drawn through a tube of cotton wool was sterile and was unchanged except for the removal of dust. Cotton plugs are still used by the bacteriologist as a convenient means of closing tubes and flasks to airborne contamination, yet freely admitting air.

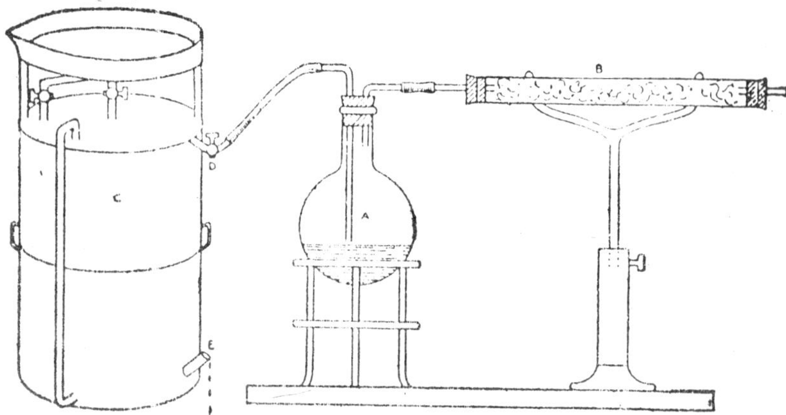

Fig. 3.1 Device used by Schroeder and Dusch for filtering air through cotton.

The final French proponents of spontaneous generation were forced into silence by the demonstration of Pasteur's that sterile broth could be left open to the air for years without contamination if the neck of the flask was drawn out and bent downward (**Fig. 3.2**). Pasteur's lone challenger, Pouchet, declined the opportunity to refute Pasteur's experiments at a public demonstration in 1864, and thus the theory of spontaneous generation lost its support in France. The use of a loosely fitting plastic or metal cap to prevent contamination of a tube or flask is an application of Pasteur's simple experiment.

Fig. 3.2 Pasteur's flasks, constructed to admit air but exclude micro-organisms.

The English proponents of spontaneous generation were more persistent than Pouchet, and it was not until after a series of experiments by Tyndall (1877), who demonstrated the heat resistance of bacterial spores, that the theory was generally discredited in England.

After it had been demonstrated that a sterilized nutrient solution could be kept sterile indefinitely, the isolation of pure cultures of bacteria needed only the development of techniques and the use of the proper media for the growth of the desired species, Robert Koch (1843-1910), a meticulous German bacteriologist, was responsible for a great part of this new methodology, and his contributions will be mentioned frequently in this chapter.

Culture Media

Bacteria were first cultured upon natural materials— pieces of potato, egg white, and meat or infusions made by boiling these materials in water. Milk was also found to be a good medium for bacteriological cultures.

Thousands of different kinds of culture media have been devised to fit the requirements of bacteria and the idiosyncrasies of the bacteriologist. In order to be of value a culture medium must support the species to be grown on it by supplying water, mineral salts, an energy source, and materials for the synthesis of bacterial cells. Additional materials called "growth factors" are also required by many bacterial species. In some instances these are the same as some of the vitamins required by animals; or they may be organic compounds of varying complexity, some of which are of unknown structure.

Culture media are generally divided into two classes—liquid and solid.

Liquid Media : Most liquid culture media are referred to as "broths," although they may only be solutions of mineral salts. This is because the first commonly used liquid culture medium was beef broth. The most commonly used liquid medium at present is *nutrient broth*. It is prepared by dissolving 5 g of peptone and 3 g of beef extract or yeast extract in a liter of distilled water. Peptone is a water-soluble material prepared by the partial enzymic digestion of a protein such as lean meat, casein or soy beans. Many varieties of peptone have been developed by varying the protein, the proteolytic enzyme (pepsin, trypsin or papain) and the time and conditions used for the digestion. Several peptones have become highly standardized products and are routinely used for special variations of culture media. In general peptones supply the majority of the nitrogenous materials needed by bacteria for the synthesis of cellular material. They may also serve as the energy source for bacterial growth. Beef extract is the water-soluble portion of beef concentrated to a paste. Yeast extract is the water-soluble portion of autolyzed yeast. They supply salts and a varitely of materials that appear to be valuable for the growth of bacteria. The use of these materials has simplified the preparation of standardized culture media, but they have not completely replaced infusions prepared from fresh meat, liver, brain, fresh vegetables, and a variety of materials., Nutrient broth and infusions are frequently modified by the addition of carbohydrates, pH indicators, and other materials used to support the growth of bacteria having special requirements or to demonstrate a physiological reaction useful in taxonomy.

Skimmed milk was one of the first fluid media used for culturing bacteria. It is still frequently used in the classification of bacteria, but is generally modified by the addition of a pH indicator such as litmus or bromocresol purple.

Solid Media : Solid culture media were first prepared by sterilizing pieces of bread or vegetables. Such material still have some use in the bacteriological laboratory, but they have generally been replaced by fluid media modified through the addition of solidifying agents such as agar, certain proteins or silica gel.

For the majority of bacteriological culture work it is advantageous to have a solid medium that can be liquified by heating and will become solid again upon cooling. The first material to be added to broths to give them this property was gelatin. Koch found that 2.5 to 5 per cent gelatin would produce a clear culture medium which was solid below 23°C Such a medium, however, could not be incubated at body temperature without becoming liquid and the gelatin was frequently destroyed by protcolytic bacteria, yielding a medium which would not gel even at a low temperature.

The use of agar to solidify culture media is attributed to a suggestion made by Frau Hesse, the wife of one of Koch's co-workers who was having difficulty with gelatin-digesting bacteria turning his solid medium into a liquid. The first mention of agar media is made in a paper published by Koch in 1882. This complex polysaccharide material derived from certain marine algae had been used by cooks as a thickening agent long before it was used in bacteriology. The addition of 12 to 15 g of agar per liter of a liquid culture medium will produce a gel that melts at about 95°C but does not solidify until cooled to below 45°C. Thus, bacteria can be introduced into a melted agar medium at a temperature that is not high enough to cause death during a short exposure, and once the medium has been solidified by further cooling, it can be incubated at temperatures as high as 60 to 70°C without melting again. Furthermore agar is not rapidly decomposed by the bacteria commonly encountered in the bacteriological laboratory. Agar has most of the properties that a bacteriologist would incorporate if he were to make a gelling agent especially for use in bacteriology.

Media containing 15 g per liter of agar are sufficiently solid to prevent the movement of most bacteria through the gel. Media with smaller amounts of agar (3 to 5 g per liter) are only semisolid and permit movement of motile bacteria while restricting the spread of non-motile bacteria by convection currents. Such media are frequently used to determine whether bacteria are motile.

Solidification of culture media with an inorganic material can be accomplished by the forming of a silicic acid gel. Such media are seldom used routinely in bacteriology because of the time required to prepare the gel and to dialyze it free of unwanted salts left from the preparation procedure. The formation of a silicic acid gel is not a process reversible by heat.

Besides gelatin, which has already been mentioned, other proteins such as egg albumin and blood serum are used to solidify bacteriological culture media. These materials can be coagulated with heat, thus producing a solid medium. Once solidified, however, they must be used in solid form. Like gelatin, they have the disadvantage of being liquefied by proteolytic bacteria and are therefore of limited usefulness.

Definition of pH, and the pH Scale : Since most bacteria are rather sensitive to changes in the acidity or alkalinity of their environment, it is frequently necessary to adjust the pH— i.e., the degree of acidity or alkalinity of the media in which micro–organisms are to be grown. A solution is acidic because of an excess of positively charged hydrogen ions (H^+) or alkaline because of because of an excess of negatively charged hydroxyl ions (OH^-). Pure water is considered neutral because when it ionizes slightly at 22°C there are equal numbers of hydrogen and hydroxyl ions liberated, i.e., $HOH \leftrightarrows H^+ + OH^-$. When a strong acid such as hydrochloric ionizes, H^+ ions are given off, i.e., $HCL \rightarrow H^+ + Cl^-$; while a strong base such as sodium hydroxide liberates OH^- ions, i.e., $NaOH \rightarrow Na^+ + OH^-$. Because of the slight ionization of water, an acid will always have some hydroxyl ions and a base some hydrogen ions present, but in both cases the relative concentrations of these ions will be small. Therefore both the acidity and alkalinity of a solution can be expressed in the terms of hydrogen ions, and for this purpose the pH scale is used. When pure distilled water ionizes at 22°C there is 0.000,000,1 gram of hydrogen ions present per liter. This may be written as 10^{-7} g per liter. For every H^+ ion there is a corresponding hydroxyl ion; thus the solution is neutral. Since the number of H^+ and OH^- ions are equal in the case of pure water, each must have a concentration of 1×10^{-7} and the product of these two ions is equal to 1×10^{-14}. On this basis a pH scale of 1 to 14 has been developed and pH has been defined as the logarithm of the reciprocal of the hydrogen ion concentration, i.e.,

$$pH = \log \frac{1}{[H^+]} \text{ or } -\log [H^+].$$

For convenience only the exponent is employed in expressing pH. As a result a solution with large numbers of hydrogen ions will have pH of less than 7, indicating an acid reaction; the presence of few hydrogen ions will result in an alkaline reaction expressed by a pH above 7.

Table 3.1 indicates the relationship of hydrogen ion concentration to pH· Note that the pH increases as the hydrogen ion concentration decreases, because the reciprocal is involved. In other words, the more hydrogen ions present, the lower the pH or the more acid the solution. Since a logarithmic scale of ten is used, a change of 1 in pH on the scale indicates a tenfold change of hydrogen ions. Hence a pH of 3 will be ten times as acid as a pH of 4 and 100 times as acid as a pH of 5.

Measurement of pH : The pH of media and solutions can be determined *colourimetrically* by means of test papers or chemical indicator solutions. With the latter, several drops of indicator are added to the medium and the colour developing based on the hydrogen ion activity is compared with a known standard. Various types of comparator blocks are available of this purpose. The indicators listed here are often used in introductory work in bacteriology.

Indicator	*P^H Range*	*Colour Changes*
Bromcresol purple	5.2-6.8	Yellow to purple
Bromthymol blue	6.0-7.6	Yellow to blue
Phenol red	6.8-8.4	Yellow to red

Table–3.1 The Relationship of Hydrogen Ion Concentration to pH

Grams of H+ per liter	***Logarithms of H + concentrations***	***Expressed as pH***	***Reaction***
1.0	0	0	Acid
0.1	–1	1	"
0.01	–2	2	"
0.00,1	–3	3	"
0.000,1	–4	4	"
0.000,01	–5	5	"
0.000,001	–6	6	"
0.000,000,1	–7	7	Neutral
0.000,000,01	–8	8	Alkaline
0.000,000,001	–9	9	"
0.000,000,000,1	–10	10	"
0.000,000,000,01	–11	11	"
0.000,000,000,001	–12	12	"
0.000,000,000,000,1	–13	13	"
0.000,000,000,000,01	–14	14	"

A second and more accurate method of determining pH is by a pH meter which permits an *electrometric* measurement. This is a more elaborate procedure than colourimetric method. It requires expensive equipment, which is first standardized using solutions of known pH and then employed to measure the hydrogen ion activity of the material being investigated.

With these methods, the pH of many substances besides bacteriological media has been determined. The following are examples of common foods with pH ranging from very acid to neutral:

Lemons	2.4	Spinach	5.4
Apples	3.4	Milk	6.6
Tomatoes	4.3	Shrimp	7.0

As for bacteria, all have an optimum range in which they grow and function best. For many this will be near the neutral point or in a pH range of 6 to 8. Others can live in or even require very acid environments such as a pH of 3 or less, whereas some prefer an alkaline pH of 8 or above. An important phase of bacterial physiology is the determination of the optimum, minimum, and maximum pH for organisms being investigated.

Specialized Media : In a modern bacteriological laboratory the growth of many bacteria in pure culture requires the use of culture media enriched with materials such as yeast extract, tomato juice, blood serum, or sterile red blood cells. The last two materials are altered by heat and must be added aseptically to agar medium that has been melted and cooled to below 50°C. On the other hand, autotrophic bacteria and also many saprophytic species can be grown on culture media of known chemical composition called *synthetic media*. The synthetic media used to grow autotrophic bacteria are merely solutions of a mixture of inorganic salt, whereas those used for the cultivation of saprophytic bacteria may range from an inorganic salt mixture plus a carbohydrate to complex mixtures of salts, carbohydrates, amino acids, vitamins, purines and pyrimidines. Increased knowledge of the physiology of bacteria and the availability of amino acids and growth factors have made it possible cultivate many bacteria on synthetic media that could formerly be grown only on media containing blood or milk. Synthetic media are useful in studying the nutritional requirements of bacteria and in some of the biological assays for vitamins and amino acids. They may also be of differential value for two organisms morphologically similar. Thus we find that *Aerobacter aerogenes* will grow will grow on Koser's citrate broth, which furnishes citric acid and sodium ammonium phosphate respectively as sources of energy and nitrogen; whereas *Escherichia coli* will not grow on this medium unless the vitamin riboflavin is provided.

Dehydrated Media : It was once a common procedure for each laboratory to prepare all of its own culture media from the basic ingredients. This practice has largely been replaced by the use of dehydrated culture media to which it is necessary to add only distilled water. Such preparations are generally stable and well standardized with respect to concentrations of materials and pH. They are of particular value in the preparation of small quantities of complex media. Many laboratories find it economical to purchase some or all of their media prepared, sterilized, and sealed in tubes or poured into disposable petri plates. Small tablets of culture media or impregnated paper discs which are merely placed in a tube of sterile water or on the surface of a basal agar medium are sometimes convenient for determining physiological characteristics of bacteria, especially in a small clinical laboratory.

Sterilization of Culture Media, Glassware, Etc

Culture media must be sterilized as soon as possible after preparation; otherwise there will be a rapid multiplication of contaminating organisms and the composition will be altered. The tubes, bottles, or flasks of media are usually stoppered with cotton, plastic sponge, slip-on metal caps or screw caps to prevent contamination after sterilization.

Methods for Sterilizing Media : Dating from Spallanzani's work during the controversy over spontaneous generation, it was customary for a time to sterilize media by heating for several hours at the *boiling point*. This method is still used in some home canning but rarely in bacteriological work because of the time involved and the uncertainty of results. It was superseded by the method of discontinuous or *intermittent heating* known as Tyndallization. The process was developed by the English physicist John Tyndall (1877) during his experiments designed to discredit the theory of spontaneous generation. As soon as tubed, the medium was heated in a steam sterilizer at the boiling point for a few minutes only. This would kill the vegetative cells but not the spores. The medium was then allowed to stand until the next day, during which time most of the spores would germinate and produce vegetative cells. A second similar heating would kill them. Since some spores would be slow in germinating, a third heating was performed on the third day. Although this method was slow and not always successful because of delayed spore germination and the changes caused by growth between heating, it was used almost universally for some years, and it is still used for certain kinds of media that are easily decomposed by excessive heat. However, approximately 8-hour intervals between heatings are generally more satisfactory than one-day periods. The method is dependent upon spore germination and can therefore be used successfully only with nutrient solutions which will support bacterial growth. It is not satisfactory for sterilized water and many solutions.

The third method, which has largely replaced the other two, is heating in an *autoclave,* or steam pressure sterilizer. The medium is placed in the autoclave the door is closed, and steam under pressure is admitted. All air must be displaced, otherwise the pressure in the autoclave will not all be due to the presence of steam. A mixture of air and steam has a temperature lower than that of pure steam at the same pressure. Therefore, in order to obtain enough heat to destroy all vegetative cells and spores it is important to remove all of the air from the chamber at the start of the sterilizing period. On all modern pressure sterilizers this feature is automatically taken care of, and a thermometer in the discharge outlet indicates the temperature in the coolest part of the chamber. The only function of the pressure developed is to increase the temperature to a point that will insure destruction of bacterial spores. The pressure most commonly used is 15 pounds per square inch in excess of the atmospheric pressure, which gives a temperature of 121°C (250°F) at sea level. This temperature is usually maintained for 15-20 minutes, although the time required for sterilization may have to be varied because if excessive loading of the autoclave or the use of large containers in which the contents are slow to heat. At the end of the sterilizing period the discharge valve and the steam valve are closed and the autoclave is allowed to cool slowly. This prevents violent boiling within the tubes or flasks, which would tend to wet the cotton plugs or even to blow them out.

Certain ingredients used in culture media, particularly some growth factors, gelatin, and some sugars, are gradually decomposed by the heat in the autoclave; and if this method of sterilization is used, the media containing them must be heated for the shortest possible time required for sterilization and cooled immediately when taken from the autoclave.

Filtration is a method that is used extensively for sterilizing solutions that may be decomposed or altered by excessive heat. A number of bacteriological and serological filters, such as the Berkefeld and Mandler, which are made of diatomaceous earth, the Chamberland and Seals, which are of unglazed porcelain, and the fritted glass filter made of powered and fused glass are available for this purpose. All are obtainable in different sizes and degrees of porosity. The efficiency of these filters in removing bacteria from the solution depends upon the size of the pores, the charge of the filter, the type of solution to be filtered, and the pressure and time employed. They are not effective in removing viruses of filterable forms of bacteria. They are prepared for use by first washing and then incinerating or autoclaving. Another type is the Seitz filter, which utilizes specially made asbestos and paper pads fitted within metal holders in which they are autoclaved before using. These pads are discarded after use. The Millipore filter consisting of a thin cellulose derivative membrane with uniform holes of subbacterial diameter, has found many uses in the field of bacteriology. Among these are the sterilization of water and blood serum.

Chemical Sterilization : Some culture media and a variety of other materials may be sterilized by chemical means. For some items it may be adequate to dip them into a chemical solution or to expose them to formaldehyde vapour. Other materials such as soil, hospital dressings, and instruments may be sterilized in a chamber containing ehtylene oxide or beta-propiolactone vapour. These gases must be handled with special equipment to avoid toxicity and explosion hazards to the personnel using them.

Sterilization of Glassware : Though the autoclave may be used for glassware, it is not to be recommended, since the glass is left wet. More commonly the tubes, flasks, Petri dishes, etc., are sterilized in a hot-air oven, using a procedure very similar to that introduced by Lister in 1878. The temperature required will depend upon the length of time during which the heat is applied. A temperature of 170°C or above for 1-2 hours is commonly used. Temperatures above 180°C will char paper or cotton. Some dry sterilizers are heated with electricity; others with gas. Uniform heating of the contents is often dependent upon proper loading of the oven.

Inoculating Culture Media

The transfer of bacteria to sterile culture media from stock cultures or from other materials containing them requires good technique and careful manipulation. At the start it must be assumed that all exposed surfaces— work tables, hands, instruments, and the exteriors of tubes and flasks of sterile media— are contaminated with miscellaneous bacteria and mold spores. Numerous micro–organisms may be borne by dust particles in the air, especially if there is a draft. The problem is to transfer the desired bacteria to the sterile culture medium without introducing any foreign ones from these sources of contamination.

Preparation of Room : Transfers are best made in a small, clean, culture room free from drafts. The air in a small space with wet surfaces soon becomes free from suspended particles of dust, which adhere to the moist areas. Therefore bacteriologists frequently work over a wet towel spread on a table or over an area which has been wiped with a wet cloth or disinfectant.

Laboratories used for very exacting work, as in the preparation of vaccines, serums, etc., are provided with filtered air which is sometimes treated with germicidal ultraviolet rays to maintain a sterile atmosphere.

Transfer with Inoculating Needle : When the exact amount of inoculum to be transferred is not an important consideration, the easiest method is by the use of an inoculating needle. The wire may be straight or the tip may be bent into a loop 1 to 3 mm or more in diameter. For large inocula several loops in the wire may be made.

In making transfers from tube to tube by this method the needle is heated in a flame to dull redness before and after each use and as much of the handle as will be thrust into the tube in making the transfer is passed quickly through the flame to burn off adhering dust.

Transfer with Pipette : For quantitative work a sterile graduated pipette is used for making transfers. For example, in making a quantitative estimate of the number of bacteria in water or milk, 1.0 ml of the sample is transferred with a pipette to a 99-ml bottle of sterile water for purposes of dilution. If it is anticipated that the number of bacteria will be very high, further dilutions are similarly made. Then 1.0 ml of the diluted sample is transferred to a sterile petri dish, and approximately 10 ml of melted agar are added and thoroughly mixed with the sample. After the agar has hardened, the dishes are inverted to prevent water that has condensed on the cover from dropping onto the medium. Excessive surface moisture encourages the development of spreading colonies which make isolations and counting of individual colonies difficult. The plates are then incubated until colonies have developed. It was once assumed that each bacterium in the sample would produce a colony, and that a count of the colonies would represent the number of bacteria in the sample; however, it is now known that, for several reasons there will be fewer colonies than there were bacteria.

Methods for Isolating Pure Cultures

Pure cultures of bacteria are rarely found in nature. As a rule, where one species is found others are found also, making a mixture that is sometimes difficult to separate. Even in the body of an animal or plant afflicted with a bacterial disease, harmless saprophytes may accompany the pathogens as "secondary invaders."

Bacteria are so small that the pioneer workers found great difficulty in separating one kind from another. As a result they often worked with mixed cultures, which they sometimes thought were pure. Various ingenious methods have been devised to assure culture purity.

The Dilution Method : Lister and his contemporaries used serial dilutions as a means of separating a single species from a mixture in which it was the predominating type. A little of the material containing a mixture of bacteria was added to one of several flasks containing a sterile liquid culture medium. From this flask a drop or so of well-mixed contents was added to other flasks, and from these to still others until, by dilution, some flasks would receive only one cell and others none at all. It might so happen that some of the flasks would receive only one kind of bacteria. One of these might chance to be of the desired kind, especially if this kind were more numerous than others in the original materials. The method was tedious and often unsuccessful. Lister's use of a finely graduated syringe to inoculate sterile solutions gave him better success than other workers had, and he is usually given the credit for having isolated the first pure culture, *Streptococcus lactis,* which he obtained from sour milk in 1878. Lister's equipment was quite crude, consisting chiefly of liqueur glasses of sterile milk covered with glass cups.

The Poured Plate Method : Among the early bacteriologists one of the most ingenious in devising methods was Robert Koch. Koch noted little specks and masses growing on the moist surface of a piece of boiled potato. Microscopic examination showed these to be, in many cases, organisms of one kind only. Potato was often an unsatisfactory medium, and in searching for an imporvement Koch was led to add ordinary household gelatin to beef bouillon as a hardening agent. To this nutrient gelatin, melted by heat, he would add a little material containing a mixture of bacteria, pour it out on a sterile glass plate, cover it with a bell jar for protection against micro-organisms from the air, and allow it cool and thereby harden. In due time colonies would appear that were sometimes pure cultures. While Koch was not the first to use solid media, or to solidify liquid media with gelatin, it was he who perfected these methods and brought them into general use.

This technique was epoch-making and has since been adopted almost universally. Two important improvements have been added. The original plates permitted the melted gelatin to flow over the edge, and they occupied too much space if used in large numbers. To remedy these defects one of Koch's students, R. J. Petri, devised shallow glass dishes with overlapping covers. These are the "petri dishes" of today. The suggestion of Frau Hesse that agar be substituted for gelatin resulted in a medium that had the advantages of not being so easily melted by warmth or liquefied by micro–organisms. These discoveries were given to the world during the period from 1883 to 1887 and are still used with very little modification.

Although pure cultures have been obtained thousands of times by this method, it happens, not infrequently, that a colony contains growths from cells of two species that happened to be together in the culture medium. Repeated plating of suspensions made from newly formed colonies usually results in purification of such mixtures but not always.

Streaking Method : As the name implies, in this method some of the inoculum is streaked or spread over the surface of a solidified medium in a petri dish. If diluted sufficiently by this technique, well-isolated colonies are obtained. This is rapid method which requires less equipment than the previous method and is commonly used for the examination of clinical specimens for the presence of pathogenic bacteria.

Poured plate and streaked plate methods can both be used for the isolation of anaerobic bacteria if the plates are incubated in a container from which oxygen is flushed with another gas or from which oxygen is removed by chemical means. Although there are many chemical reactions which consume oxygen, the combustion of hydrogen is the one which is least likely to produce toxic products or absorb carbon dioxide which may be required for growth. Special containers are available for carrying out this reaction with a minimum of explosion hazard.

If tubes of agar overlaid with paraffin to exclude air are substituted for petri plates, isolated colonies of anaerobic bacteria can be obtained in the depths of the culture medium. Such colonies, however, may be more difficult to transfer to other media than those grown in a poured plate incubated under anaerobic conditions.

Single Cell Isolation : The ideal way to obtain a pure culture from a mixture of micro–organisms would be to pick out individual cells of the kind wanted, just as one might pick out one kind of garden seed from a mixture of several. The difficulty with this procedure is that bacteria are so small that they cannot be seen with the unaided eye, and instruments fine enough to pick up a single one are difficult to make and use.

These difficulties have been in a measure overcome, however, by a combination of a micro-pipette and the micro-manipulator. The pipette, with a very fine capillary point, is mounted on the micro-manipulator, which is an instrument with micrometer adjustments by means of which the pipette or other instrument can be moved forward and backward, right and left, and up and down.

In a simple but effective procedure a series of hanging drops of diluted culture are placed on a special sterile cover slip by a micro-pipette which has a bent tip. A second pipette is inserted from below into one of the hanging drops which has been observed to contain only a single bacterium. The cell can be drawn into the pipette by gentle suction and then transferred to a large drop of sterile broth on another sterile cover slip. After incubating in a hanging drop slide until the cell has multiplied through several generations, sufficient cells are present to insure growth when a dropful is transferred to a tube of broth. Although this appears to be the most accurate method for obtaining pure cultures, it is not generally used for the isolation of bacteria because the micro-manipulator is expensive, its manipulation is very tedious, and there are numerous difficulties appreciated only by those who have used the instrument. This method is better adapted for the isolation of yeasts and molds.

Animal Inoculation : A method occasionally used for obtaining pure cultures of pathogenic organisms is the inoculating of the mixture containing them into a susceptible animal. The intruding saprophytes will not multiply within the body of the animal but will usually be destroyed, while the pathogen will multiply and may then be transferred to culture media. This is the method commonly used for the isolation of *Mycobacterium tuberculosis* from the sputum of persons suspected of having tuberculosis. Similarly, pure cultures of the pneumococcus may often be obtained by inoculating sputum containing the pneumonia organism into a white mouse. From such animals the organism can be transferred to tubes of culture media during *post mortem* examination. Koch (1878) believed animal inoculation to be the best and surest method of obtaining pure cultures of pathogenic bacteria.

Animals are also used extensively in the preparation of vaccines and immunizing sera. In addition they play an important role in the cultivation of viruses, rickettsiae, and *Treponema pallidum*, the causative agent of syphilis, for which they serve as a culture medium. In recent years they have been employed widely in determining the toxicity of new drugs and antibiotics before they are tested on man.

Stock Culture : Every important bacteriological laboratory now keeps pure cultures of those organisms that it needs either for student use or for research. Such supplies are called *stock cultures* and are kept with meticulous care. Each species is grown on a medium that has been found to be favourable for its growth and survival. Standard nutrient agar, in test tubes plugged with cotton, is suitable for most saprophytes, most plants pathogens, and some human and animal pathogens.

It was once the custom for bacteriologists to supply stock cultures on request to their colleagues in other institutions, but this practice became burdensome and was often unsatisfactory in its results because of delay and sometimes incorrectly identified or contaminated products. For this reason the American Society for Microbiology was instrumental in establishing the American Type Culture Collection of bacteria, fungi, and some viruses. Similar culture collections are maintained in other countries.

Stock cultures of most species keep best when refrigerated. They are then practically dormant, so that injurious products of metabolism accumulate but slowly. Only a few species fail to survive at refrigeration temperature, Even when preserved by this method, it is generally necessary to transfer stock cultures at intervals of a few months.

In a later chapter the harmful effects of low temperatures and desiccation on bacteria will be discussed. However, it is interesting to note at this point that these two factors are now frequently employed in the preservation of stock cultures. A common procedure is to suspend the bacteria in sterile skim milk or serum and then transfer 0.1 ml quantities to small vials. A thin film is frozen over the inside surface of the vial by rotating it in a mixture of alcohol and solid carbon dioxide at a temperature of –78°C. The vials are immediately evacuated. This dries the bacteria while they are still frozen. Finally the tubes are sealed *in vacuo* with a small flame. These cultures can then be stored for several years at 4°C. Even pathogenic bacteria have been held for several years without loos of morphological or physiological characteristics, including the ability to produce diseases. To revive such cultures it is merely necessary to break open the vial aseptically, add a suitable sterile medium, and after incubation make further transfers. This drying process, known as *lyophilization*, permits one to maintain large numbers of cultures without the variation induced by repeated transfers. Also the danger of contamination is greatly decreased.

The principle of freeze-drying has been applied to many areas other than that of preserving bacterial cultures. Blood plasma and serum are preserved in this manner; and bones arteries, and skin now can be conveniently banked for future use. Food-processing industries are finding increasing applications for freeze-drying to provide food products which, upon reconstitution with water, have the flavour and nutrient value of the natural materials.

Cultural Characteristics

Pure cultures of bacteria can be studied by growing them on a wide variety of culture media. The production of pigment is best observed on solid media. Also many species make growths quite characteristic with regard to shape, texture, contour of surface, luster, etc., when cultivated on these media. These growths are studied mostly with the unaided eye or with simple lens of low magnification. The three commonest ways of preparing such growths are (1) by inoculating the slanting surface of a tube of the medium with a stroke of an inoculating needle; (2) by a stab into the depths of a translucent tube of culture medium, usually nutrient agar or gelatin; and (3) by mixing a few of the bacteria with agar or gelatin, previously melted by heat and partially cooled, and allowing the mixture to solidify in a petri dish, where colonies will form, each being the result of the multiplication of an individual.

Bacterial Colonies : The bacteriologist is afforded many opportunities for observing colonies of bacteria. Petri dishes are used for estimating the number of bacteria in milk, water, soils, etc., as well as for obtaining pure cultures. Thus, incidental to the main purpose for which the plating is done, the colonies come under observation. For more precise studies of colonies, plates are inoculated from pure cultures, A few species, such as *Sarcina lutea* and *Bacillus cereus* var. *mycoides*, can be recognized with fair certainty by the appearance of their colonies; but most species are not distinctive, i.e., colonies of different species often look much the same. The colony appearance common to the greatest number of species is grayish white, round, smooth, glistening, slightly convex, and 3 to 5 mm in diameter. Dozens of species produce colonies that fit this general description.

The characteristics of surface colonies that receive most attention are:

1. *Size.* This may vary from a small fraction of a millimeter (pinpoint colonies) to the entire surface of a petri dish (spreaders).
2. *Shape.* The commonest shape of colonies is circular, but others have scalloped or lobed margins, and these lobes may be extended and branched like the leaves of a fern.
3. *Elevation.* The colonies may be thin and flat, or thickened after different patterns.
4. *Contour.* The surface may be even and smooth or wrinkled or may show ridges or point-like elevations.

4

The Microscope

Bacteria and viruses are so small that they cannot be seen with the naked eye. They must be greatly magnified before they can be clearly seen and studied. The use of a microscope is, therefore, absolutely indispensable to the bacteriologist and to the biologist in general.

A microscope may be defined as an optical instrument, consisting of a lens or a combination of lenses, for making enlarged or magnified images of minute objects. The term is compounded from the two Greek words *μῖκρός* micro, small, and *σκοπεῖν* scope, to view.

A simple microscope, or a single microscope, consists merely of a single lens of magnifying glass held in a frame, usually adjustable, and often provided with a stand for conveniently holding the object to be viewed and a mirror for reflecting the light. A compound microscope differs from a simple one in that it consists of two sets of lenses, one known as an objective and the other as an eyepiece, commonly mounted in a holder known as a body tube (Fig. 4.1). The one nearest the specimen, called the objective, magnifies the specimen a definite amount. The second lens system, the eyepiece, further magnifies the image formed by the objective, so that the image seen by the eye has a magnification equal to the product of the magnifications of the two systems. The individual or initial magnification of the objectives and eyepieces is engraved on each such part. Accurate focusing is attained by a special screw appliance known as a fine adjustment. Compound microscope give much greater magnifications than simple microscopes and are necessary for viewing and examining such minute objects as bacteria.

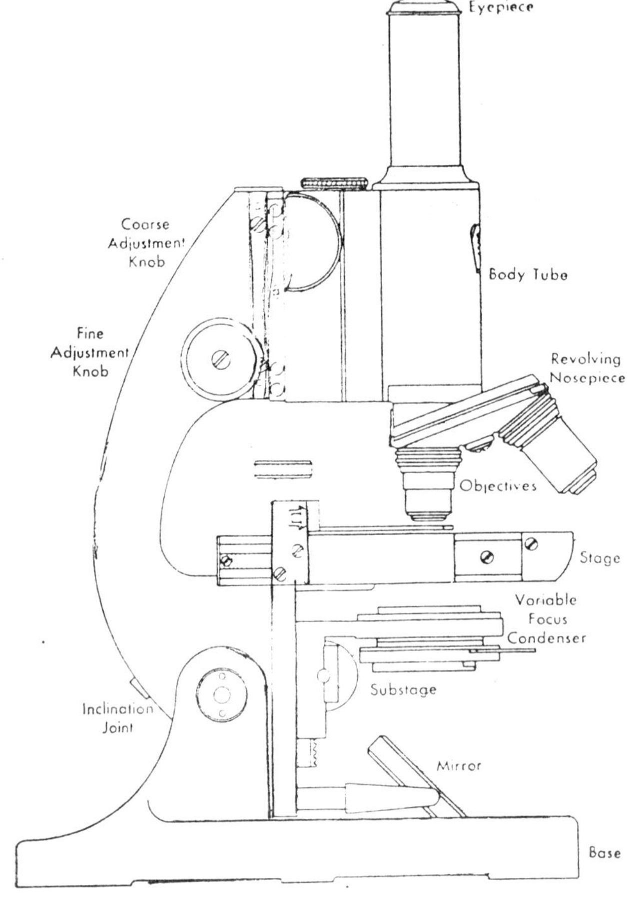

Fig. 4.1 Compound microscope and its parts

Every user of the microscope should first understand the principles involved in order that the instrument may be employed to the greatest advantage. As Sir A. E. Wright stated:

Every one who has to use the microscope must decide for himself the question as to whether he will do so in accordance with a system of rule of thumb, or whether he will seek to supersede this by a system of reasoned action based upon a study of his instrument and a consideration of the scientific principle of microscopical technique.

General Principles of Optics

The path of light through a compound microscope is illustrated in Fig. 4.2. The light, in passing through the condenser, object in plane I, and objective lens, would from a real

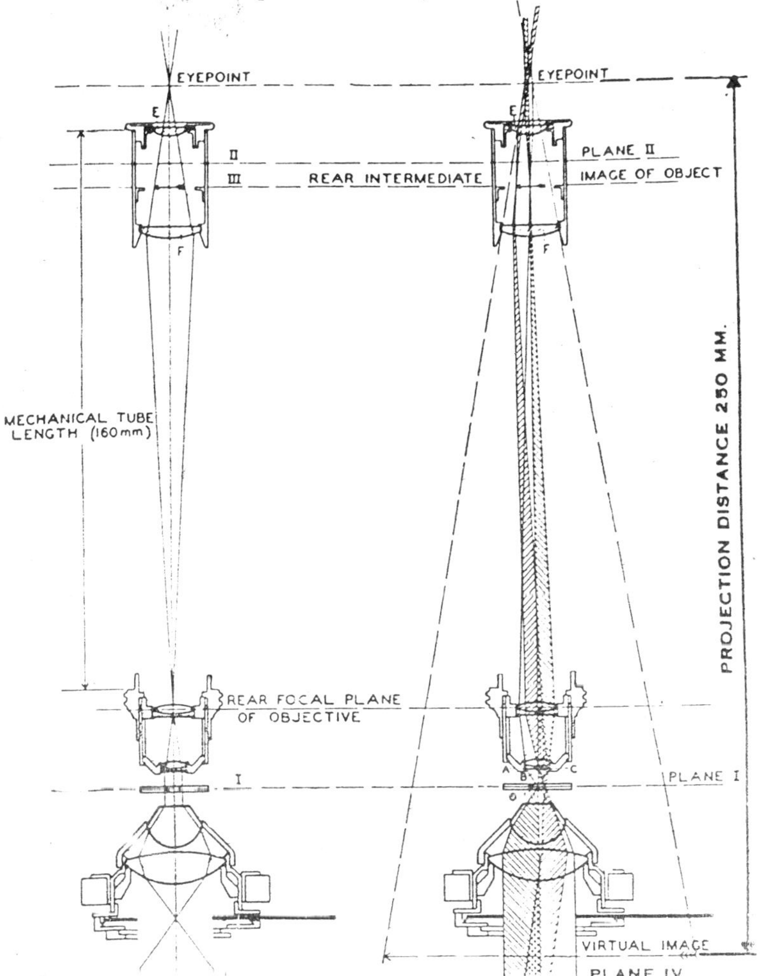

Fig. 4.2 Path of light through a microscope.

and inverted image in plane II if the ocular or eyepiece were removed. In the presence of the ocular F, the rays are intercepted, forming the image in plane III. The real image is then examined, with the eye lens E of the ocular acting as a single magnifier and forming a virtual image in Plane IV. The distance between the virtual image (plane IV) and the eyepoint is known as the projection distance. The object is magnified first by the objective lens and second by the ocular, or eyepiece. With a tube length of 160 mm. (Most microscope manufacturers have adopted 160 mm. As the standard tube length), the total magnification of the microscope is equal to the magnifying power of the objective lens multiplied by the magnifying power of the ocular.

The above magnifications are obtained on a ground glass placed 10 in. From the ocular of the microscope. After the microscope has been set at the proper tube length, the total magnification may be computed by multiplying the magnifying power of the objective by that of the eyepiece and by one-tenth of the distance from the eyepiece to the ground glass measured in inches. For example, if the ground glass is placed 10 in. from the eyepiece of the microscope, the total magnification will be as given on the ocular and objective. If the ground glass is placed 20 in. from the eyepiece, the magnification will be twice as great. If placed 5 in. From the eyepiece, the magnification will be one-half as great. To take a specific example:

Magnification of objective......................	97 ×
Magnification of ocular..........................	10 ×
Distance of ground glass from ocular.....	7 in.
Total magnification.................................	97 × 10 × 0.7 (0.1 × 7) = 679 ×

It may be seen that almost any degree of magnification could be obtained by using oculars different magnifying powers or by varying the length of the draw tube. Even though the magnifying powers of the microscope could be greatly increased in this manner, the amount of detail that can be seen is not improved since this is strictly limited by the structure of light.

Structure of Light : According to the undulatory, or wave, theory, light is transmitted from luminous bodies to the eye and other objects by an undulatory or vibrational movement. The velocity of this transmission is about 186,300 miles per second, and the vibrations of the ether are transverse to the direction of propagation of the wave motion. The waves vary in length from 3850 to 7600 angstrom units (A.) approximately. The colour evoked when the energy impinges on the retime varies in a complex way with the wave length, amplitude character whose lengths fall above or below the limits mentioned are not perceptible to the average eye under normal conditions. Those between 1000 and 3850 A. Constitute ultraviolet light and are manifested by their photographic or other chemical action. Those exceeding 7600 A. are the infrared waves and are detected by their thermal effects.

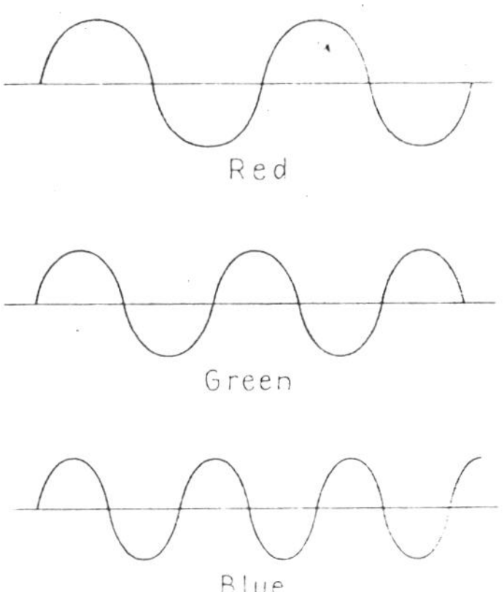

Fig. 4.3 Wave lengths of light of different colors.

When a beam of white light is passed through a prism, a spectrum is obtained in which several colours form a series from deep red through orange, yellow, green, blue, and indigo to deepest violet. The wave lengths of the various colours are different; red shows the longest and violet the shortest waves of the visible spectrum.

The length of a light wave is the distance from the crest of one wave to the crest of the next (**Fig. 4.3**). The unit of measurement is the angstrom unit (A.), which is equal to 1/10,000,000 mm., or to approximately 1/250,000,000, in. The visible spectrum, together with the corresponding wave lengths of the light rays in angstrom units, may be represented as shown in Fig.4.4. Visible light waves, ranging in length from 4000 to

Blue Violet	Blue Green	Green	Orange Yellow	Red

4000 5000 6000 7000

Fig. 4.4: Light rays of the visible spectrum and their corresponding wave lengths in angstrom units.

7000 A., may be roughly divided into three portions; blue-violet, from 4000 to 5000 A.; green, from 5000 to 6000 A.; red, from 6000 to 7000 A.

Objectives

The objective is the most important lens on a microscope because its properties may make or mar the final image. An objective capable of utilizing a large angular cone of light coming from the specimen will have better resolving power than an objective limited to a smaller cone of light. The chief functions of the objective lens are (1) to gather light rays coming from any point of the object, (2) to unite the light in a point of the image, and (3) to magnify the image.

There are three major types of objectives, namely, achromatic, fluorite, and apochromatic. The achromats are the simplest in construction and the least expensive. They are adequate for most purposes, Correction for both colour and spherical aberration is quite good in the lower-power objectives, but the control of aberrations becomes more difficult as the power is increased. Aberrations are largely eliminated by the use of fluorite (semiapochromatic) objectives and, especially, the apochromats. The latter are more highly corrected with respect to aberrations than any other type of objective and are preferred for the most critical work.

Numerical Aperture : The resolving power of an objective may be defined as its ability to separate distinctly two small elements in the structure of an object that are a short distance apart. The measure for the resolving power of an objective is the numerical aperture (N. A.). The larger the numerical aperture, the greater the resolving power of the objective and the finer the detail it can reveal.

Since the limit of detail or resolving power of an objective is fixed by the structure of light, objects smaller than the smallest wave length of visible light cannot be seen. In order to see such minute objects, it would be necessary to use rays of shorter wave length. Invisible rays, such as ultraviolet light, are shorter than visible rays but since they cannot be used for visual observation (photography only), their usefulness is limited.

The image of an object formed by the passage of light through a microscope will not be a point but, in consequence of the diffraction of the light at the diaphragm, will take the form of a bright disk surrounded by concentric dark and light rings. The brightness of the central disk will be greatest in the center, diminishing rapidly toward the edge. The image cone of light composed of a bright disk surrounded by concentric dark and light rings is spoken of as the antipoint. If two independent points in the object are equidistant from the microscope lens, each will produce a disk image with its surrounding series of concentric, dark and light rings. The disks will be clearly visible if completely separated, but if the images overlap they will merge into a single bright area, the central portion of which appears quite uniform. The two disks will not, therefore, be seen as separate images. It is not known how close the centers of the images can be and still be seen as separate antipoints.

The minimum distance between the images of two distinct object points decreases as the angle of light *AOC*, coming from the object, *O*, increase. The angle formed by the extreme rays is known as the aperture of the objective. The ability of the objective lens system to form distinct images of two separate object points is proportional to the trigonometric sine of the angle. The latter, then, is a measure of the resolving power of the objective. Actually, however, the sine of angle *AOB* is used, which is just one-half of angle *AOC*. This is usually referred to as sin μ. Since the sine of an angle may be defined as the ratio of the side opposite the angle in a right-angled triangle to the hypotenuse, then

$$\sin \mu = \frac{AB}{AO}$$

The light in passing through the objective is influenced by the refractive index n of the space directly in front of the lens. This is another factor that affects the resolving power of an objective. The two factors, refractive index n and sin μ, may be combined into a single expression, the numerical aperture, which may be expressed as follows:

Importance of N. A : If a very narrow pencil of light is used for illumination, the finest detail that can be revealed by a microscope with sufficient magnification is equal to

$$\frac{w.l.}{N.A.}$$

Where w. l. is the wave length of the light used for illumination and N. A. is the numerical aperture of the objective. The resolving power of the objective is proportional to the width of the pencil of light used for illumination. This means that the wider the pencil of light, the greater the resolving power. The maximum is reached when the whole aperture of the objective is filled with light. In this instance, the resolving power is twice as great. The finest detail that the objective can reveal is now equal to

$$\frac{w.l.}{2N.A.}$$

For example, the brightest part of the spectrum shows a wave length of 5300 A. An objective having a numerical aperture equal to 1.00 will resolve two lines separated by a distance of 5300 A./ 1.00 = 5300 A. (48,000 lines to the inch) if a very narrow pencil of light is used, and 5300 A./ (2 × 1.00) = 2650 A. 95,000 lines to the inch) if the whole aperture of the objective is filled with light.

From the above, it is evident that the maximum efficiency of an objective is not reached unless the back lens is filled with light. This may be ascertained by removing the eyepiece from the microscope and viewing the back lens of the objective with the naked eye. If the back lens is completely filled with light, the efficiency will then be according to the numbers engraved on the objective.

Resolving Power : The shorter the wave length of light, the finer the detail revealed by the objective. With an objective having an N. A. Of 1.00 and a yellow filter (light transmission of 5790 to 5770 A.), it is possible to see about 88,000 lines to an inch; with a green filter (light transmission of 5460 A.), about

95,000 lines to an inch; with a violet filter (light transmission of 4360 A.), about 115,000 lines to an inch; and with ultraviolet light (light transmission of 3650 A.), about 140,000 lines to an inch.

Immersion Objectives : When a dry objective is used, an air space is present on both sides of the microscope slide and cover slip. The largest cone of light coming from *O* (**Fig. 4.5**) that could possibly be used is 180° in air, which is equal to an angle of about 82° in the glass. This corresponds to a numerical aperture 1.0. In actual practice, however, these figures become 143 and 77° respectively, owing to the fact that the air space must be wide enough to correspond to a practical working distance of the objective. Rays of greater angular aperture than 82° in glass, which originate at the object point *O* by diffraction, will be completely reflected at the upper surface of the cover slip *t*.

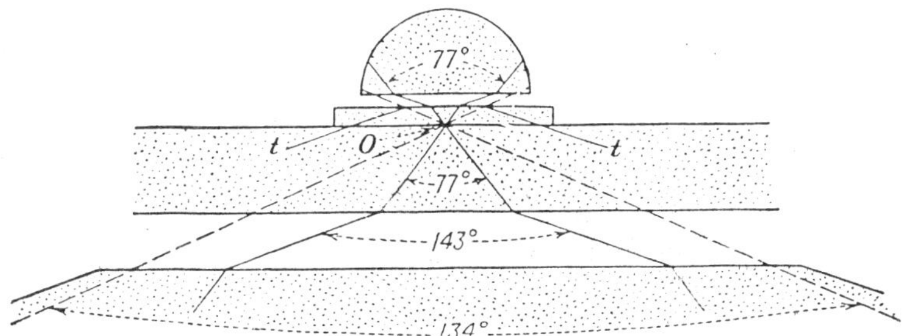

Fig. 4.5 Passage of light through an object on a glass slide using dry and immersion objectives. See text for details.

The refractive index *n* of the air is equal to 1.0. If the air space between the cover slip and the objective is filled with a fluid having a higher refractive index, such as water (n = 1.33), or, what is still better, a liquid having a refractive index approaching that of glass, such as cedarwood oil (n = 1.51), angles greater than 82° are obtained. Numerical apertures greater than 1.0 are realized by this method. Cedarwood oil causes the light ray to pass right through the homogeneous medium, with the result that a cone of light of about 134° is obtained, which corresponds to a numerical aperture of 1.4. Finer detail can, therefore, be resolved by this procedure. With an oil-immersion objective and a numerical aperture of 1.4, two lines as close to ether as 1/100,000 in. (0.2 μ) can be separated. This means, then, that the greater the numerical aperture of the objective, the greater will be its resolving power or ability to record fine detail.

The refractive indexes of a number of media that have been employed for immersion objectives are given in Table 4.1.

Table 4.1

Medium	*Refractive index at 25°C*
Water	1.33
Mineral (Paraffin) oil	1.47
Cedarwood oil	1.51
Sandalwood oil	1.51
Shillaber's immersion oil	1.52
Balsam	1.53
Crown oil	1.55

Depth of Focus : The depth of focus is known also as the depth of sharpness or penetration. The depth of focus of an objective depends upon the N. A. and the magnification, and is inversely proportional to both. This means that the higher the N.A. and the magnification, the lower the depth of focus. Therefore, high-power objectives must be more carefully focused than low-power objectives. These conditions cannot be changed by the optician.

Equivalent Focus : Objectives are sometimes designated by their equivalent focal lengths measured in either inches or millimeters. An objective designated by an equivalent focus of $\frac{1}{12}$ in., or 2 mm., means that the llens system prodces a real image of the object of the same size as is produced by a simple biconvex or converging lens having a focal distance of $\frac{1}{12}$ in., or 2 mm.

Working Distances of Uncovered Objects : If the object on a glass slide is not covered with a cover lip, the working distance may be defined as the distance between the front lens of the objective and the object on the slide when in sharp focus. The

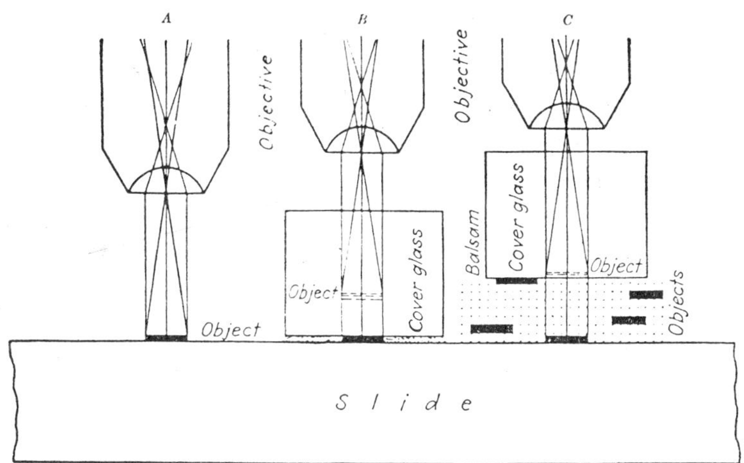

Fig. 4.6 Working distance of an objective. *A*, object not covered with a cover slip. *B*, *C*, object covered with a cover slip.

working distance is always less than the equivalent focus of the objective.

The working distance may be determined easily by noting the number of complete turns of the micrometer screw (fine adjustment) required to raise the objective from the surface of the slide, where the object is located, to a point where the microscope is in sharp focus.

To take a specific example :

Each turn of the micrometer screw = 0.1 mm,
Number of turns required to bring object in sharp focus = 6
Working distance = 6 × 0.1 = 0.6 mm.

Working Distance of Covered Objects : If the object is covered with a cover slip, the free distance from the upper surface of the cover slip to the front of the objective will be less than in the case of an uncovered object. It is obvious from this that if the cover glass is thicker than the working distance of the objective, it will be impossible to get the object in focus. On the other hand, if the glass is thin it will be possible to get the object in focus, but the focus of the microscope on a covered object will be different from that on an uncovered object. It follows from this that an object covered with a glass cover slip or other highly refractive body will appear as if raised, and the amount of elevation will depend upon the refractive index of the glass or other medium covering the object. Also the greater the refraction of the covering body, the more will be the apparent elevation (See Fig. 4.6 B, C). The apparent depth of the object below the surface of the covering medium may be calculated by taking the reciprocal of its index of refraction. For example, if a glass cover slip is used, it will have an index of refraction of 1.52. The reciprocal of this figure is 1/1.52 = 2/3, approximately. This means that the apparent depth of the object is only two-thirds its actual depth.

The working distance of covered objects may be determined by noting the number of complete turns of the micrometer screw fine adjustment) required to raise the objective from the surface of the cover slip to a point where the objective is in sharp focus.

To take a specific example:

Each turn of micrometer screw = 0.1 mm.
Number of turns required to bring object in sharp focus = 3.5
Working distance = 3.5 × 0.1 = 0.35 mm.

Aberrations in Objectives : Perfect lens systems have not yet been designed. All lens systems have aberrations to a greater or lesser degree, depending upon the skill of the designer and the magnitude of the design problem. Lens systems are made up of lenses having spherical surfaces, and such surfaces do not form perfect images. This defect may be largely counteracted by combining lens shapes and different glasses.

The principal defects in the image are the result of chromatic aberration, spherical aberration, distortion, curvature of field, astigmatism, coma, and lateral colour.

Chromatic Aberration : White light in passing through a prism is broken up into its constituent colours, the wave lengths of which are different. A simple or compound lens, composed of only one material, will exhibit different focal lengths for the various constituents of white light. This is due to the dispersive power of the lens. Every wave length is differently refracted, the shortest waves most and the longest waves least. The blue-violet rays cross the lens axis first and the red rays last. There will be a series of coloured foci of the various constituents of white light extending along the axis (Fig. 4.7). As a result, the lens will not produce a sharp image with white light. Instead, the image will be surrounded by coloured zones or halos which interfere with the visual observation of its true colour. This is spoken of a chromatic aberration. It may be lessened by reducing the aperture of the lens, or better still, by using a lens composed of more than one material (compound lens). Two or more different glasses or minerals are necessary for correcting the chromatic aberration of an objective, and the amount of correction depends upon the dispersive powers of the components of the objective.

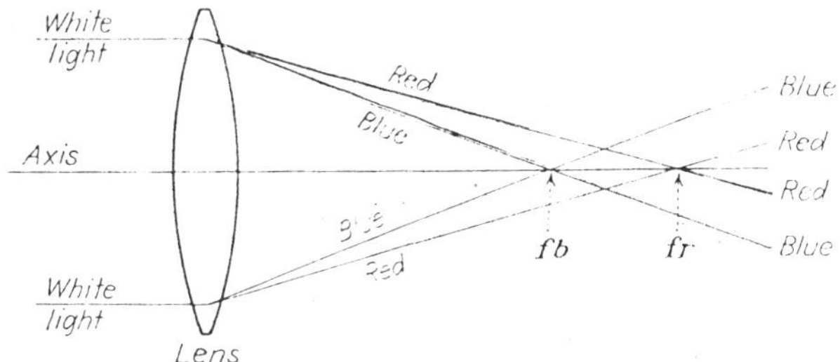

Fig. 4.7 Chromatic aberration with white light. White light, in passing through a lens, is dispersed into its constituent colours. The red or long waves are refracted less than the blue or short waves. The blue rays (*fb*) cross the optical axis of the lens before the red rays (*fr*). The blue light will focus nearer the lens than the red light.

If two optical glasses are carefully selected to image light of two different wave lengths at the same focal point, the lens is said to be achromatic, and an objective containing such a lens system is spoken of as an achromatic objective. The remaining rays of the white light will be imaged at approximately the same point. An achromatic objective will yield images free from pronounced colour halos. If the focus is shifted slightly, faint green and pink halos may be observed. The slight residual colour will not prove objectionable for the usual microscopic work. Achromats are the universal objectives for visual work and are very satisfactory in photomicrography when used in monochromatic light (obtained by the use of filters).

Lens systems corrected for light of three different wave lengths are called apochromatic objectives. These objectives are composed of fluorite in combination with lenses of optical glass.

The images produced by objectives in this group exhibit only a faint blue or yellow residual colour. Since these objectives are corrected for three colours instead of for two, they are superior to the achromats. Their finer colour correction makes possible a greater usable numerical aperture. The violet rays are brought to the same focus as visual rays. This fact makes these objectives excellent for photographic use for both white and monochromatic light.

Another group of objectives exhibit qualities intermediate between the acrhomats and the apochromats. They are called *semiapochromats*. If the mineral fluorite is used in their construction, they are termed fluorite objectives. These objectives also yield excellent results when used for photomicrography.

Spherical Aberration : This refers to the greater power in the outer portion of a spherical surface than in the inner portion. This is overcome by judicious combinations of convergent and divergent lens elements, properly shaped to minimize the variation of focal power with aperture. Spherical aberration causes some of the light which should be in the central spot to diffuse out into the ring structure. This causes a loss in contrast in the normal microscope preparation.

Distortion : This type of aberration renders a square object as an image with curved sides. If the rulings near the edge appear curved inward, it is known as cushion distortion. If the opposite effect occurs, where the rulings appear curved outward, it is known as barrel distortion. Distortion is caused by the lens surface having different magnifications at the marginal and central portions of the image.

Curvature of Field : This aberration is caused by a spherical lens surface which produces a curved image of a flat object due to the marginal portions of the image coming to a focus at a different distance than the central portions of the image

Astigmatism : If a marginal point object is drawn out into two separate line images lying at different distances from the lens surface, it is called astigmatism. It results in a general deterioration of the off-axis image. An astigmatic image can never be focused sharply except for detail parallel or perpendicular to a radius of the field.

Coma : This name is given to the defect in which different circular concentric zones of the lens surface give different magnifications to an off-axis image. This results in a point object being imaged as a comet-shaped image. Coma in the center of the field is an indication of damage to the objective.

Lateral Colour : The presence of this defect results in light of one colour being imaged at a greater magnification than light of another colour. This causes an off-axis image of a point object to be spread out into a tiny spectrum or spread of colour.

Oculars

The chief functions of the ocular, or eyepiece, are the following:

1. It magnifies the real images of the object as formed by the objective;
2. It corrects some of the defects of the objective;
3. It images cross hairs, scales, or other objects located in the eyepiece.

Several types of eyepiece are employed, depending upon the kind of objectives located on the microscope. Those most commonly used are known as Huygenian, hyperplane, and compensating oculars.

Huygenian Eyepiece : In this type of eyepiece, two simple plano-convex lenses are employed, one of which is below the image plane (Fig. 4.8). The convex surfaces of both lenses face downward. Oculars in this group are sometimes spoken of as negative eyepieces. This type of ocular is made with a large field lens, which bends the pencils of light coming from the objective toward the axis without altering to any great extend the convergence or divergence of the rays in the individual pencils. Above the field lens, and at some distance from it, is a smaller lens known as the eye lens, the functions of which is to convert each pencil of light into a parallel or only slightly diverging ray system capable of being focused by the eye. The rays, after emerging through this lens, then pass through a small circular area known as the Ramsden disk, or eyepoint. It may be seen that the real image of the object is formed between to two

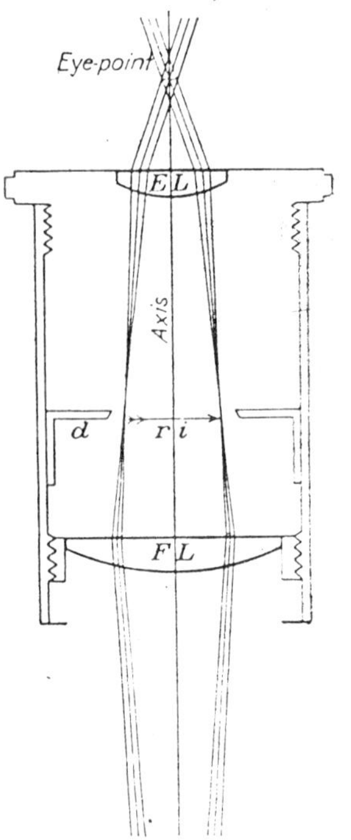

Fig. 4.8 Huygenian eyepiece. *EL*, eye lens; *FL*, field lens; *RI*, real images formed between the ocular lenses and the diaphragm d.

eyepiece lenses. In an eyepiece of this type, the distance separating the two lenses is always a little greater than the focal length of the eye lens. The reason for this is to prevent any dirt on the field lens from being seen sharply focused by the eye. An image should be viewed with the eye placed at the Ramsden disk in order to obtain the largest field of view sand also to obtain the maximum brightness over the field.

The Huygenian eyepiece works well with the low-power achromats but gives under-corrected curvature of field and lateral colour with the intermediate and higher power objectives. The degree of compensation required increases with the objective power, making it highly desirable to have a graded series of eyepieces. Therefore, the Huygenian eyepiece should be used to cover the low powers; the hyperplane eyepiece, the intermediate powers; and the compensating eyepiece, the high powers.

Hyperplane Eyepiece : Apochromatic objectives, when used with compensating eyepieces, give fields that are not flat. Flat-field eyepieces have been designed to correct this defect. They give much flatter fields than do the other two types but they are lless perfectly corrected chromatically. Oculars of this type are referred to as hyperplane, planoscopic, periplane, etc. They may be employed with the higher power achromatic, fluorite, and apochromatic objectives without introducing chromatic aberrations in the image. Their colour com-pensation falls about midway between the Huygenian and the compensating eyepieces.

Compensating Eyepiece : Oculars of this type consist of an achromatic triplet combination of lenses (**Fig. 4.9**). These eyepieces are more perfectly corrected than are those of the Huygenian and hyperplane types. A compensating eyepiece is corrected to neutralize the chromatic difference of magni-fication of the apochromatic objectives. Such eyepieces are intended, therefore, to be used primarily with apochromatic objectives, although they may be employed with the higher power achromatic and fluorite objectives with good results.

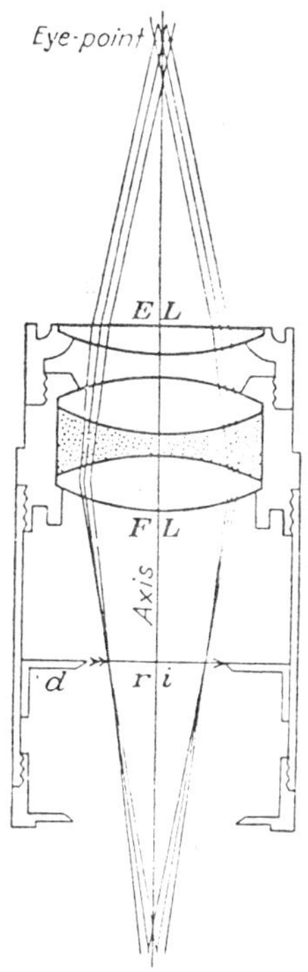

Fig. 4.9 Compensating eyepiece. *EL*, eye lens; *FL*, field lens composed of three components, *RI*, real image formed below the lenses at the diaphragm d.

Condensers

Several methods are employed for illuminating the object under examination. In bacteriology, the two methods commonly used are (1) illumination by transmitted light and (2) dark-field illumination.

Illumination by Transmitted Light : A condenser may be defined as a series of lenses for illuminating, with transmitted light, an object to be studied on the stage of the microscope. It is located under the stage of the microscope between the mirror and the object, whereas the objective and ocular lenses are located above the stage. It is sometimes referred to as a substage condenser. The most popular substage optical system is known as the Abbe condenser (**Fig. 4.10**).

A condenser is necessary for the examination of an object with an oil-immersion objective to obtain adequate illumination. A condenser is also preferable when working with high-power dry objectives. Probably the most commonly employed condenser has a 1.25 N. A.

A good condenser sends light through the object under an angle sufficiently large to fill the aperture of the back lens of the objective. When this is accomplished, the objective will show its highest numerical aperture. This may be determined by first focusing the oil-immersion objective on the object. The eyepiece is then removed from the ocular tube. The back lens of the objective is observed by looking down the microscope tube, care being taken not to disturb the focus. The back lens of the objective should be evenly illuminated. If it in not, the mirror should be properly centered. If the condenser has a smaller numerical aperture than the objective, the peripheral portion of the back lens of the objective will not be illuminated, even though the condenser iris diaphragm is wide open. If the condenser has a greater numerical aperture than that of the objective, the back lens of the objective may receive too much light, resulting in a decrease in contrast. The smaller the aperture, the greater the depth of focus and the greater the contrast of the components of the image. The lowest permissible aperture is reached when diffraction bands become evident about the border of the object imaged. This difficulty may be largely overcome by closing the iris diaphragm of the condenser until the leaves of the iris appear around the edges of the back lens of the objective. The diaphragm is then said to be properly set. The setting of the iris diaphragm will vary with different objectives.

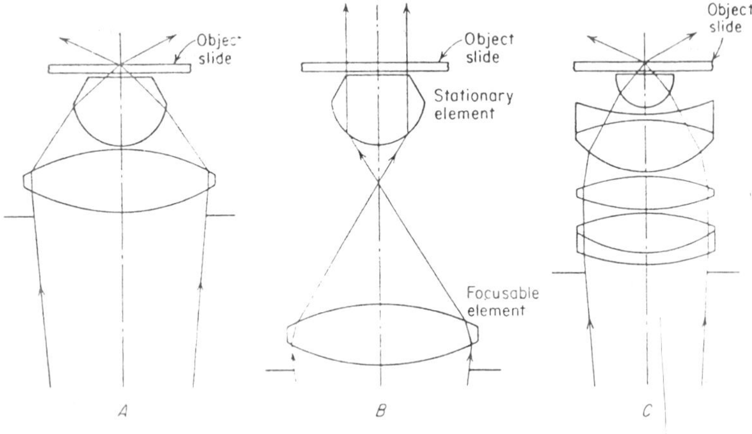

Fig. 4.10 Substage condensers. A. Abbe Condenser; B, variable-focus condenser; C, achromatic condenser. (Courtesy of Bausch & Lomb Optical Company.)

The Abbe condenser is a 1.25- N. A. condenser utilizing only two lenses. Because of its simplicity and good light-gathering ability, it has become extensively used for general microscopy. It is, of course, not corrected for spherical or chromatic aberration, but for general visual observation it serves very well.

The variable-focus condenser (**fig. 4.10** *B*) is a two-lens condenser, 1.25 N. A. maximum, in which the upper lens element is fixed and the lower one focusable. By this means it is possible to fill the field of low-power objectives without the necessity of removing the top element. This condenser is basically similar to the 1.25 N. A. Abble when the lower lens is raised to its top position. When the focusable lens is lowered, the focus of the light is brought in between the elements, and when this focus is at the point indicated in the diagram, the light emerges as a large-diameter parallel bundle.

The achromatic condenser (**fig. 4.10** *C*) is a 1.40- N.A. Condenser that is corrected for both chromatic and spherical aberrations. Because of its high degree of correction, it is recommended for research microscopy and colour photomicrography where the highest degree of perfection in the image is desired.

Dark-field Illumination : The microscope is most commonly employed by allowing the light to pass through the object. This is called microscopy in transmitted light or bright-field microscopy. An object cannot be seen in bright-field microscopy unless it absorbed or refracts the light passing through it. Contrast is thus set up between the object and the surrounding medium. Objects that display feeble contrast with the background are difficult to see in bright- field illumination.

If the aperture of the condenser is opened completely and a dark-field stop inserted below the condenser, the light rays reaching the object form a hollow cone. If a stop of suitable size is selected, all the direct rays from the condenser can be made to pass outside the objective. Any object within this beam of light will reflect some light into the objective and be visible. This method of illuminating an object, where the object appears self-luminous against a dark field, is known as dark-field illumination.

Three types of condensers are employed for dark-field illumination: (1) the Abbe, (2) the paraboloid, and (3) the cardioid.

The Abbe condenser is probably more commonly employed than the other two because it is especially suitable for objects that do not require the highest magnifications to make them visible. It may be employed either by inserting a dark-field stop below the condenser (fig. 4.11 *A*) or by unscrewing the top part of the condenser and substituting for it a dark-field element (fig. 4.11 *B*).

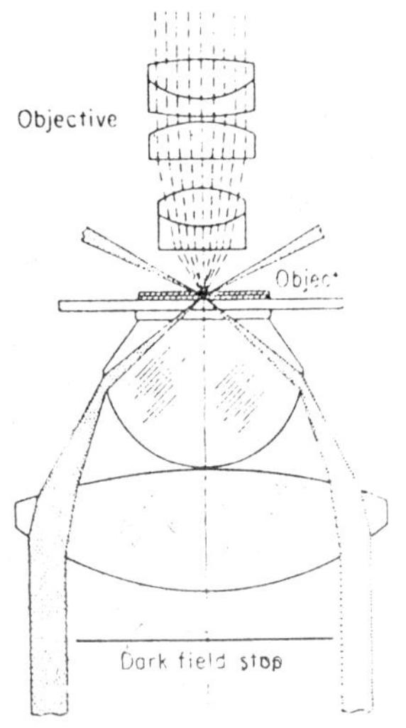

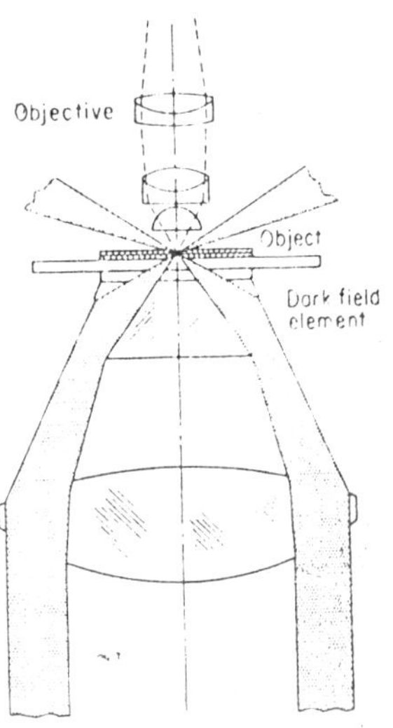

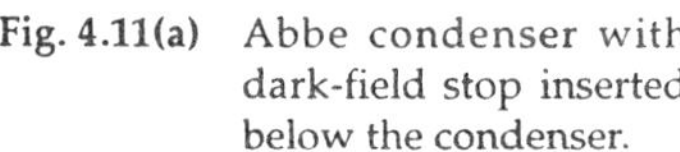

Fig. 4.11(a) Abbe condenser with dark-field stop inserted below the condenser.

Fig. 4.11(b) Abbe condenser.

The paraboloid condenser is designed to be used with high-power oil-immersion objectives and an intense source of light (fig. 4.12). In using this condenser, it is necessary to place cedar oil or glycerin between the condenser an the slide. Also, the specimen must be mounted in a liquid or cement and protected with a cover slip. The numerical aperture of the objective must not be greater than that of the condenser.

The cardioid condenser is the most refined type of dark-field illuminator (fig. 4.13). It is especially designed to be used for the examination of colloidal solutions or suspensions, i.e., particles measuring less than 0.25 μ in diameter.

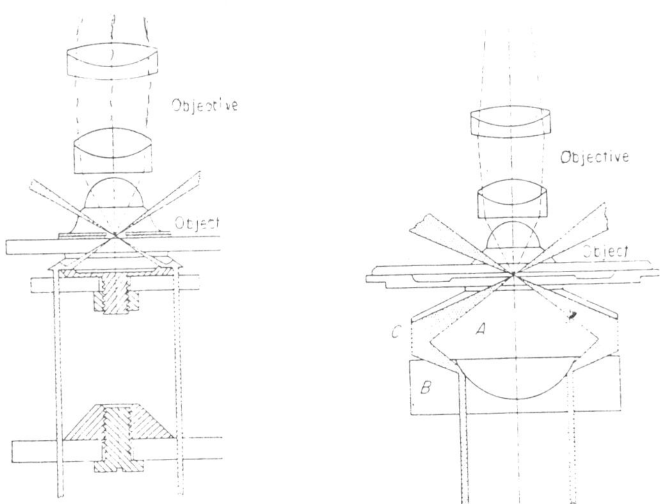

Fig. 4.12 Paraboloid condenser. **Fig. 4.13** Cardioid condenser.

The cardioid condenser is best employed with a strong are lamp. Since the concentration of light is so great, ordinary glass slides and cover slips should not be used. Visible defects and the difficulty of removing foreign objects from the glass ruin the visibility of ultramicroscopic particles. It is better to employ fused-quartz object slides and fused-quartz cover slips. They are free from bubbles and other imperfections and can be heated in a flame to drive off all dirt, after being chemically cleaned. Pijper (1951) recommended the use of thin microscope slides to which were cemented pieces of mica of similar shape by means of Canada balsam. When the balsam had hardened, a thin layer of mica was split off with a sharp knife, which left an untouched, scratch-free, and dust-free surface, perfectly suited for dark-field microscopy.

For more information on the microscope see Allen (1940), American Optical Company (1945), Gage (1941), Muñoz and Charipper (1943), and Wredden (1948).

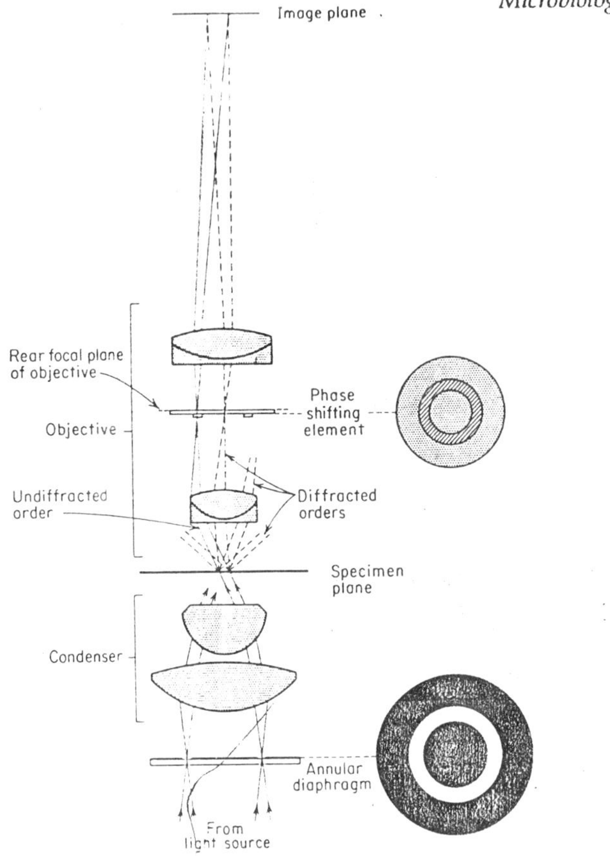

Fig. 4.14 Schematic diagram of the general optical arrangment of a phase microscope.

Phase Microscopy

The principle of the phase microscope is not new, having been discovered as early as 1892, but it has only recently been developed to practical use. It complements rather than replaces existing methods of microscopy.

Phase microscopy is a method for controlling the contrast in the image and making visible unstained living micro-organisms and cytological details within them. Phase microscopy also greatly enhances visibility of stained material of low contrast.

The method employed for fixing and staining bacteria may make a difference in their size. The bacterial cell usually shrinks considerably during drying and fixing. Since micro–organisms show sharp edges under the phase microscope in unstained preparations, accurate measurement of living cells is now possible.

A schematic diagram of the general optical arrangement of a phase microscope is shown in (Fig. 4.14). An annular aperture in the diaphragm, placed in the focal plane of the substage condenser, controls the illumination on the object. The aperture is imaged by the condenser and objective at the rear focal plane, or exit pupil, of the objective. In this plane a phase shifting element, or phase plate, is placed.

Light, shown by the solid lines and undeviated by the object structure, in passing through the phase altering pattern, acquires a one-quarter wave length of green light advance over that diffracted by the object structure, (broken lines) and passing through that region of the phase plate not covered by the altering pattern. The resultant interference effects of the two portions of light form the final image. Altered phase relations in the illuminating rays, induced by otherwise invisible elements in the specimen, are translated into brightness differences by the phase altering plate.

Fluorescence Microscopy

Fluorescence is a property of some substances to excite emission of visible light by the absorption of invisible ultraviolet radiations. The exact mechanism of this phenomenon is not clearly understood. Some materials are autofluorescent, whereas others can be made to fluoresce by treatment with fluorescent chemicals called fluorochromes. Equipment includes a source of ultraviolet rays, suitable filters for isolating the radiations, and a standard microscope.

Some applications of fluorescence microscopy include examination of acid-fast bacteria (*M. Tuberculosis*, etc.), differentiation of living and dead micro–organisms, examination of hair for mold spores, and location of chemotherapeutic agents in tissues.

Electron Microscopy

As already mentioned, particles less than half the wave length of the light employed in the microscope system cannot be resolved. This barrier to the microscopic examination of the majority of the viruses has been overcome by the use of the electron microscope. This instrument, developed over the past

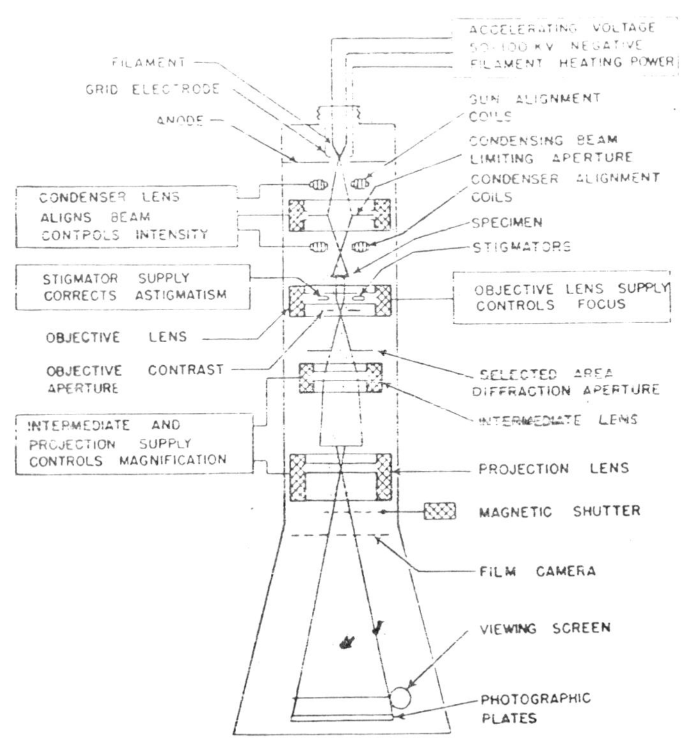

Fig. 4.15 RCA electron microscope, Type EMU-3. Note general principles of construction.

20 years, employs electrons, which have a wave length of 0.05 Å, instead of light. The theoretic limit of resolution of the electron microscope is there fore very low, but in practice, commercial models give resolution of 10 Å or better.

Although the simpler technics of electron microscopy have become standardized, considerable experience is still required before operators can obtain consistently good results from commercial instruments. Machines manufactured by the Radio Corporation of America, Siemens, and Philips are those best known to workers in North America and Europe.

In general terms, as shown in figure 4.15, electrons are generated from a tungsten filament and pass down the microscope tube, which is kept evacuated. These electrons are next condensed by an electromagnetic lens and pass through the specimen. In the specimen, dense material scatters or absorbs the electrons, which otherwise continue to pass down the tube.

The electrons are magnified by the objective lens, and further by the intermediate and projector lenses, and finally reach a photographic plate or fluorescent screen, Direct magnifications of up to 100,000 are obtained.

Specimens for electron microscopy are prepared in many ways. Many of the classic studies were done with preparations dried in air one collodion mounts. This process leads to some distortion of the shape of the virus particles. This has been largely replaced by a method in which the virus containing suspension is sprayed onto a carbon coated disk at the temperature of liquid nitrogen. The disk is placed in *vacuo*, and the ice sublimes when the temperature is raised.

In this freeze drying method there is much less distortion, and polyhedral instead of spherical forms are seen. The shadow-casting of air or freeze dried preparations is almost routine. A, thin layer of chromium, gold, platinum or uranium is deposited at a tangent so that the virus elementary bodies appear as it were in relief. Many electron micrographs of elementary bodies in this text are of shadowcast preparations. For purpose of measuring and counting elementary bodies, special technics have been devised, in which latex particles of known dimensions are mixed

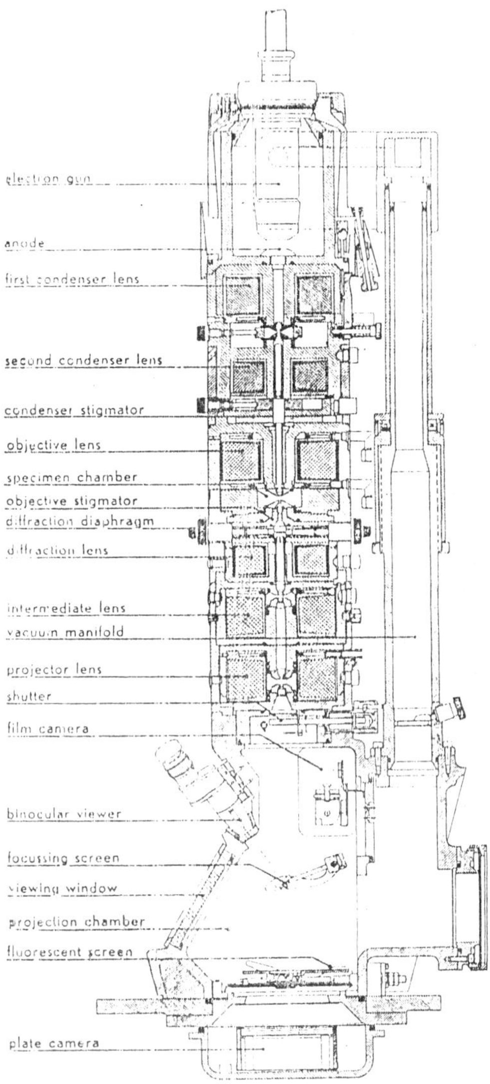

Fig. 4.16 Specifications and operational details. Philips Norelco EM-200 electron microscope.

with the virus.

Resolving Power : 10Å and better.

Magnification Range : Stepwise control, with the aid of a single panel control knob, without change of pole pieces, in 20 calibrated steps, each having a fixed magnification. 1,500-200,000 × on photographic plates. 700–90,000 × on 35 mm, film. Continuous control—which can be switched on at any time—by using several combinations of lenses, without change of pole pieces and specimen holder—from 300 × (for screening to 500, 000 × on 6½ in. Fluorescent screen.

Accelerating Voltage : 40, 60, 80 and 100 KV, variable during operation.

Micrographic Facilities : 35 mm. Film camera (40 exposures), provided with automatic exposure counter. Plate camera for 12 plates of, 3¼" × 4" or 3¼" × 3¼" provided with automatic exposure counter. Electronic exposure meter, measures the average electron intensity of the whole image. Electronically controlled shutter for time exposures as well as preset exposures from 0.5 to 60 seconds. Exposure time control and (push-button) shutter control, within easy reach of operator, on control panel. Shutter can be partially opened to register two exposures on a single plate. Stereomicrographs obtainable with standard holder, angles can be set to plus or minus six (6) degrees. Vacuum desiccating chamber for film and plates, Micron marker makes permanent reference dimensions directly on film or plate. Dark micrography using offset objective aperture.

Focussing : "Wobbler" device for rapid focussing at lower magnifications. Binocular viewer (9×) with 1 square inch ultra fine grain focussing screen, which can be moved in and out of the beam. 6 inch fluorescent screen can be tilted parallel with the viewing window, Fine focus control adjustable in steps of 0.05-0.1 - 0.2 and 0.4 microns.

Compensation of Astigmatism : Electromagnetic compensation of astigmatism on both the objective and 2nd condenser lens. Electro-magnetic stigmators are centerable; no image shift during compensation. Compensation of objective lens astigmatism to less than 0.1 micron. Compensation is possible on any specimen, simultaneously with focussing, at magnifications from 20,000 × upward.

Electron Diffraction : Transmission diffraction patters of any party of the specimen obtainable within 30 seconds, without displacing the specimen. Diaphragm, consisting of four externally adjustable platinum blades, for obtaining diffraction patterns from areas as small as 1 square micron. Holder for three standard apertures, interchangeable with normal diffraction diaphragm holder, for high-selective diffraction studies. Central beam stop.

Specimen Chamber Cooling : Norelco cooling device lowers temperature of specimen chamber to approximately —90 degrees centigrade, preventing specimen contamination and decreasing contamination of aperture.

Specimen Holder : Single specimen holder for normal work, for the entire magnification range, for stereomicrographs and for electron diffraction. Specimen holder for 3 mm, specimen grids.

Aperturing : Multiple aperture holders of similar type for first and second condenser lens, for objective lens and for diffraction lens (high resolution diffraction). Three standard disk-type apertures in each holder. Aperture holders can be externally adjusted and centered. Wide range (from 25 to 300 microns) of apertures available.

Electron Optical System (Water Cooled) : V-shaped tungsten filament with D. C. Supply, external stepwise emission control. Double condenser lens, providing minimum beam spot of 1-2 microns diameter on specimen; condenser system provided with two aperture holders, electro-magnetic stigmator and beam deflection device; change from single to double condenser lens operation and *vice versa* within 30 seconds. Objective lens provided with electro-magnetic stigmator and aperture holder and "wobbler" device. Specimen stage, with smooth and drift free spacimen translation, permitting entire bridged area specimen to be scanned. Specimen is placed between the objective lens pole pieces (Norelco's exclusive "immersion" lens). Space between pole pieces equals 5 mm. Diffraction lens (for the lower range of magnifications) provided with externally adjustable diffraction diaphragm or standard aperture holder for high-selective diffraction studies; stepwise as well as continuous current control. Intermediate lens for the upper range of magnifications; stepwise as well as continuous current control. Projector lens having continuous current control as well as fixed magnification for fully illuminated screen. Projection chamber, provided with micron marker, automatic shutter, film camera intermediate focussing screen (tiltable parallel to the viewing window 6½ in, fluorescent screen, binocular viewer (9×) and plate camera.

Stabilization : Electrical stability:

1:200,000 for lens currents. 1:100,000 for the high voltage. Guaranteed at slow line fluctuations or deviations of ±10%.

Alignment : Each lens is independently adjustable, mechanically and electrically, to permit optimum alignment. Lens current modulator can be switched on during critical centering of either condenser stigmator, objective stigmator or objective lens enabling the operator to use both hands during centering operation.

Maintenance : Swivel crane mechanism permitting rapid access to any part of optical system. Each pole piece easily removable for inspection or cleaning. Complete recentering of column is unnecessary when a single component has been re-inserted after cleaning. Unitized electronics, readily accessible, in separate cabinet. Five lens current stabilizing units interchangeable. Monitoring system for checking and measuring stabilities with built-in microvoltmeter.

Pumping System : Controlled by single, positive action, manually operated, valve. Rotary prevaccum pump (not in operating during actual use of the instrument). Buffer tank (reservoir). Mercury diffusion pump, water cooled. Oil diffusion pump, water cooled.

Evacuation Times : From cold start, 25 min. After changing filament, 5 min. After changing film camera, 10 min. After changing plate camera, 10 min. After changing specimen. 20 sec.

Water Cooling : For cooling of lenses and high vacuum pumps, 1 liter per minute required at 30p. s. i., 25 degrees C. Maximum temperature.

Power Consumption : 4.5 K.V-a, cos $\phi = 0.7$.

Weight : Cabinet, approximately 1100 Ibs. Console, approximately 1550 Ibs.

Line Voltages : 110-150, 200-250, 330-440 V. Permissible deviation ± 10% single phase 50 and 60 c/s.

One of the great advances in modern systematic virology is the negative staining technic, which is showing rich rewards in the hands of Horne, Waterson, Wildy and others. The technic is relatively simple and involves spraying virus with 1 per cent phosphotungstic acid (PTA). It is by this technic that details of the intimate substructure of virus bodies are being worked out. For example, an adenovirus particle has been shown to be an icosahedron with 252 subunits each measuring 70Å. Fascinating information about the nucleoprotein inner componenets of virus bodies is being obtained by examination of disrupted particles.

A classic method of electron microscopic examination involves the making of a replica. This technic gives beautiful portrayals of virus crystals. Briefly, a film of virus crystals is coated with carbon, and the mount and crystals are dissolved. The mould or "replica" that remains is photographed. Such preparations illustrate the phenomenon of packing symmetry, and the polygonal shape of the constituent elementary bodies can be recognized.

Another most important technic is that of ultramicrotomy, which enables very thin sections of cells or virus particles to be cut with a glass or diamond knife in an ultramicrotome. Tissue culture cells or aggregates of virus elementary bodies are prepared for sectioning by fixing in osmium tetroxide and embedding in plastic. The use of this technic has revealed many significant features of virus development. For example, the inclusions of molluscum have been shown to be divided by septa; it has been found that the elementary bodies of herpes simplex form in the nueleus, and then pass into the cytoplasm obtaining a second outer membrane; and crystals of adenovirus have been found in the nucleus of HeLa cells, the individual elementary bodies forming a lattice, the center-to-center distance between the bodies being 60 to 65 mμ.

The vigorous study of the intricacies of virus substructure by means of various electron microscopic technics is likely to throw much light on the mode of virus multiplication, and is already providing a basis for a rational system of classification based on the conventional study of morphology.

FILTRATION

Older Types of Filters

The early work of Pasteur and others on the separation of bacteria from viruses was performed with porcelain and earthenware filter candles. Among the most extensively used has been the Pasteur-Chamberland that is available in many grades (L_1 to L_{13}) according to porosity. The Berkefeld and Mandler filters are made of Kieselguhr or diatomaceous earth, and have been used for preliminary clarification.

The disadvantages of these filters are that they are liable to become clogged, and can be cleaned only with difficulty, as by trypsin or heating to high temperature, when they may develop cracks or flaws. Another objection is that they absorb large amounts of virus.

The practical value of earthenware filters is thus limited to the recovery of relatively small amounts of virus from material contaminated with bacteria. The introduction of penicillin and streptomycin for destroying such contaminants seems to have diminished still further the usefulness of these filters in virus isolation work. Sales micro-porous filter are made of pure porcelain and do not contain diatomaceous earth. They are free of some of the disadvantages of the earlier types of candle, and can be used to prepare bacteria-free suspensions of virus containing material.

A good coarse filter for withholding bacteria is the Sitz asbestos disk type as used in general bacteriology, which is less subject to the disadvantages mentioned.

Elford's Collodion Membrane Filters

Thin membranes prepared either by Elford's method or by that of Bauer and Hughes can be manufactured in a range of porosities of known average pore diameter (APD) varying from

3μ to $10m\mu$. Parlodion or other nitrocellulose is first dissolved in a mixture of alcohol, ether, and acetone. Thereafter, to measured volumes of this solution are added varying concentrations of water and acetic acid; each mixture is poured into a flat bottomed glass cell, and is allowed to evaporate for a fixed time at constant temperature and relative humidity. By addition of water, the permeability of membranes can be increased, and by addition of acetic acid decreased. Membranes are cut into small disks, and the average pore size diameter is calculated by estimating the rate of water flow at known pressure for membranes of measured thickness, surface area, dry, and wet weight. Membranes are washed with water, sterilized by boiling or ultraviolet light, and stored in distilled water. For use, a membrane is clamped in a specially designed metal holder, and a positive or negative pressure of 25 cm, of mercury is applied. The size of viruses is estimated by testing for infectivity filtrates obtained from membranes of varying APD. Table 4.2 shows the estimated diameters of many viruses arrived at by filtration. Filters of this type can be bought commercially.

Table—4.1

Relationship Between Size of Virus Elementary Bodies Determined by Electron Microscopy and Filtration through Elford's Filters

Virus	*Size determined by electron Microscopy*[a]	*Size Determined by Ultrafiltration*			*Ratio Electron Microscopy/ Filtration*
		APD[b] *passing virus*	*APD holding virus*	*Average*	
	$m\mu$	$m\mu$	$m\mu$	$m\mu$	
Influenza A	80	180	140	160	0.50
Adenovirus	80	160	140	150	0.53
Rabbit Papilloma	45	72	62	67	0.67
Coxsackie	28	50	41	46	0.61
Poliovirus	27	40	27	34	0.79
Theiler's FA	27	55	27	31	0.87

[a] Either by freeze drying technic or by measurements made in crystalline arrays.

[b] APD = average pore diameter.

Estimates of the size of virus particles obtained. by Elford' method have often been larger than the figures obtained by the more precise methods of electron microscopy. Black, in reinvestigating this problem, has shown that the ratio of the size determined by electron microscopy to that obtained by filteration varies from 0.05-0.87. He concludes that, on the average, the figure for size determined by filtration should be multiplied by factor of 0.64, to give a figure close to that determined by electron microscopy of freeze dried particles.

Centrifugation

Although virus particles are very small, they nevertheless possess measurable mass and can be deposited by centrifugal force. The ordinary bucket type of horizontal centrifuge has served its purpose in the past, but even at 10,000 revolutions per minute (r.p.m.), the rate of sedimentation is slow, and several hours may be required to produce an appreciable deposition. Angle head centrifuges have been introduced as a means of reducing the distance to be traversed by sedimenting particles.

The Spinco Preparative Ultracentrifuge

This uniquely designed and engineered centrifuge is eminently suitable for sedimentation of viruses and macromolecular proteins.

Among many valuable features, the following may be mentioned. Rotors are interchangeable, and are loaded with sealed plastic or metal tubes; speed and time of run are automatically controlled; and the rotor vacuum chamber is enclosed in armour plating and refrigerated so that low temperatures can be maintained during operation. Since the instrument is of the self balancing type, large scale preparative runs of quantities up to 1,600 ml. Of material may be handled without accurate balancing. Rotor #21, with 10 tubes, inclined at an angle of 18°, has a total capacity of 940 ml. A maximum centrifugal force of 59,000 g. At 21,000 r. p. m. Is attainable in an acceleration time of 15 minutes. Rotor #40, with 12 tubes, inclined at an angle of 26°, has a capacity of 162 ml.; a maximum centrifugal, force of 144,700 g. At 40,000 r. p. m. is attainable in an acceleration time of 5 minutes.

5

Bacterial Taxonomy

The information presented in this chapter illustrates how the morphologic, bio-chemical, immunologic, and genetic characteristics of a bacterium are used for its identification. The technique used to examine thee properties are described, and a brief description of the major group and types of bacteria that are included in the field of medical microbiology is included.

Classification of Bacteria

Nomenclature :

Like all other organisms, bacteria are given a binomial name, consisting of a genus, a species, and, if appropriate, a subspecies or serotype. These names are printed in italics; examples include organisms such a *Escherichia coli* and *Bacteroides fragilis* subsp. *Fragilis*. Based on the higher order classification used for plants and animals, this type of nomenclature implies a knowledge of evolutionary or phylogenetic relationships existing among different bacteria. Until recently, such an assumption would have been very tenuous, since no fossil records exist to provide a history of the line of descent of the various genera. Another impediment to the establishment of phylogenetic relationship among bacteria results from their rapid evolutionary change. Which causes a blurring of those key features that might be used for such a classification. Consequently, the classifications and groupings of bacteria have been based only on certain prominent morphological and biochemical features.

A major system for the classification of bacteria is presented in *Bergey's Manual of Determinative Bacteriology* which was first published in 1923 and had its latest revision in 1984. This system has avoided the use of higher phylogenetic relationships and has divided all bacteria into various parts (or hierarchies), based on prominent morphologic and metabolic characteristics. Thus, this Manual serves a guide or key for the rapid naming of a genus and species.

MORPHOLOGIC CLASSIFICATIONS

One major criterion for classification is the morphology of bacteria (i.e., their size and shape). This is routinely determined after staining the bacteria, a procedure that permits the visualization of the general shape and, in many cases, reveal additional features of cellular contents or structures that are useful for identification.

Techniques for Microscopic Study of Bacteria

Living Bacteria :

Living bacteria are difficult to see using the light microscope because individual cells appear almost colourless, even though the culture as a whole may be highly coloured. However, it is often desirable to look at living bacteria,, particularly to determine if they are motile. In these observations, one must remember that a bacterium is considered motile only if it seems to be moving in a definite direction. Even non motile bacteria bounce back and forth because of bombardment from water molecules (Brownian motion).

Wet mounts of bacteria are usually prepared by placing a drop of a liquid culture on a glass slide and covering it with a cover slip. The edges of the cover slip can be sealed with petroleum jelly to reduce evaporation of liquid and prevent convection currents.

Stained Bacteria :

Bacteria are far more frequently observed in stained smears than in the living state because the size, shape, and arrangement of cells can be much more easily perceived if the cells have been coloured by reaction with stains. Special stains are used to identify cellular structures. To prepare bacteria for staining, a small amount of culture is dispersed in a drop of water on a glass slide. The smear is allowed to dry at room temperature and the bacteria are fixed to the glass by gentle heating in a flame.

General Staining Reactions

Most stains are salts, one ion of which is coloured. In basic dyes, the cation (positive ion) is coloured, whereas in acidic dyes, the colour is carried by the anion (negative ion). Bacteria have a high affinity for basic dyes, because of the high content of negatively charged macro-molecules found in the cell wall, nucleic acids, and ribosomes. Commonly used basic dyes are crystal violet, safranin, and methylene blue. Simply covering a smear with one of these dyes for 30 to 60 seconds, followed by rinsing with water, is sufficient for showing the size and shape of the cell. When some basic dyes bind to highly negatively charged cellular constituents, they undergo a change in colour, the so-called metachromatic shift. Blue dyes, such as methylene blue, appear red when bound to nucleic acid or polyphosphate granules.

Because acidic dyes are repelled by the negative charge of the bacterial cell, dyes such as Congo red or nigrosin can be used for negative staining. Only the background will be coloured, and the colourless cell bodies will stand out. This technique is valuable for showing the shape of smaller cells.

Another form of negative staining is used for detection of the capsules around cells, which are not normally stained by the common dyes. Cells are suspended in India ink, which provides a black background in which the clear cells are visible. Because the ink particles are too large to penetrate into the capsule, cells that have capsules are seen to have a large, clear area surrounding the cell body.

Gram Stain :

An extremely important staining reaction was developed in 1884 by a Danish physician, Christian Gram. The procedure has four steps: (1) the heat-fixed smear is covered with a solution of gentian or crystal violet; (2) after 60 sec, the dye is washed off and the smear is flooded with an iodine solution; (3) after 60 sec. the iodine solution is washed off with water and the slide is rinsed with 95% alcohol for 15 to 30 sec; and (4) the slide is counter stained for 30 sec; with safranin (a red dye) or Bismarck brown.

The violet basic dye and the iodine form a large, complex aggregate within the heat-fixed cell. This complex can be readily washed out of some types of bacteria with alcohol; these are called gram-negative organisms. These appear red in the final slide because they are stained only with the safranin dye. Other bacteria retain the dye complex during the alcohol wash; these are called gram-positive organisms. They appear blue or bluish-purple because the red counterstain is not detectable on cells that have retained the intensely coloured blue complex.

The gram staining reaction of a cell is a reflection of the nature of its cell wall. The walls of gram-positive bacteria are much thicker and more cross-linked than in gram-negative cells, and it is this cell wall meshwork that traps the large dye complex. Gram-negative bacteria have much thinner, lipid-containing walls, which are partly dissolved by alcohol, allowing the dye complex to be washed out of the cell body.

Acid-Fast Stain :

The acid-fast stain is used primarily for the detection of bacteria in the genus *Mycobacterium*, which includes the agents of tuberculosis and leprosy. Mycobacteria have a very hydrophobic surface, which resists entry of dyes by the usual staining procedures. To get the dyes into these cells. Harsh conditions are used. Such as heating the organisms in the stain or including detergents in the stain. However. Once these cells are stained by carbolfuchsin (a mixture of phenol and the dye fuchsin), their cell wall structure allows them to retain the stain even when washed with 95% alcohol containing 3% HCl–hence, the term *acid-fast*. All other bacteria are decolourized by this procedure. This stain allows the detection of small numbers of acid-fast bacteria even when they are present with large numbers of other cells, as in the sputum from a patient with tuberculosis.

Other Stains :

A number of additional specialized staining procedures are used to stain parts of the bacterial cell; some examples will be seen in the micrographs throughout this text. They include techniques for staining capsules, cell walls, chromatin, flagella, endospores, and other structures. All these staining procedures involve the use of two or more special dyes, but none are used routinely in the identification of a bacterium.

Shapes of Bacterial Cells

The shapes of medically important bacteria are classified into the following general forms: (1) cocci, or spherical cells; (2) bacilli, or cylindrical or rod-shaped cells; and (3) spiral or curved forms. A few bacteria exhibit filamentous morphology consisting of branched tubes and some bacteria lack a constant shape and characteristically have a variable (pleomorphic) morphology.

Cocci :

Cocci (singular, coccus; from the Greek word for "berry") are generally spherical, although often elongated in one dimension. Cocci often exist in groups that are characteristic of a species. For example, most streptococci form long chains of cells because all progeny cells divide in the same plane and do not easily separate. *Streptococcus pneumoniae* may also occur in short chains, but it is predominantly seen in pairs. Cocci that divide in two planes to form tetrads of cells belong to the genus *Goffkya* and those that divide sequentially in three planes to form cubical packets are found in the genus *Sarcina,* Cells that divide randomly in the three planes form irregular clusters and are classified in either the genus *Staphylococcus* or *Micrococcus.*

Bacilli :

Bacilli (singular, bacillus; meaning "little staff") are rod-shaped. The ends of some bacilli are rounded, whereas those of others (fusiform bacteria) are more tapered. Some bacilli are so short that they are called *cocco-bacilli*. All bacilli divide across the narrow axis and can be seen in any arrangement from single cells to long chains. However, unlike the cocci, the length of the chains of bacilli is not an identifying characteristic.

Spiral Forms :

The spiral bacteria can be divided into three groups. Vibrios are curved rods, resembling commas. *Spirilla* are S-shaped but do not extend for more than one sine wave, whereas spirochetes are helical and flexible spiral-shaped organisms that are always longer than one wave.

Biochemical Classifications

Taxonomic classifications also make use of many prominent biochemical properties of an organism. For example, their response to oxygen and the requirement for its presence or absence during growth reflects several basic metabolic capabilities, which can be easily measured. Moreover, some medically important bacteria are exquisitely sensitive to killing by the mere presence of oxygen, and special precautions must be taken for their isolation and growth.

The ability to use specific sugars has also been a primary basis for assignment of a bacterium to a particular species in many genera. Because the measurement of growth using a sugar as sole carbon source is impractical for those bacteria having complex nutritional requirements, it is more common to test for the production of acid by such cells growing on a rich medium containing that sugar. Simple methods to test for the production of characteristic metabolic end-products such as gas and hydrogen sulfide are available. Colourimetric assays may be used to detect the presence of various enzymes which may distinguish differences among various species or genera. A good example of this is the production of high levels of urease by the genus Proteus in contrast to most other enteric bacteria.

Relatedness among a large number of organisms can be estimated by a numerical analysis in which 100 to 300 biochemical characteristics are compared for each strain. A similarity index is calculated for each pair of organisms, using the formula S = NS/(NS + ND), where S is the similarity index, NS is the number of characters common to both organisms, and ND is the number of characters that are not shared. The higher the value of S, the more related are the pair of strains. Examination of a large number of biochemical parameters reduces the bias inherent in less extensive classification schemes that are based on the presence or absence of only a few key enzymes (Fig. 5.1).

Serologic classifications are used very commonly to distinguish differences among species within a given genus. These tests employ antibodies as very sensitive and specific probes for the presence of various chemicals and antigenic conformations present on the bacterial surface. The use of monoclonal antibodies has greatly increased the specificity for many such antigenic probes.

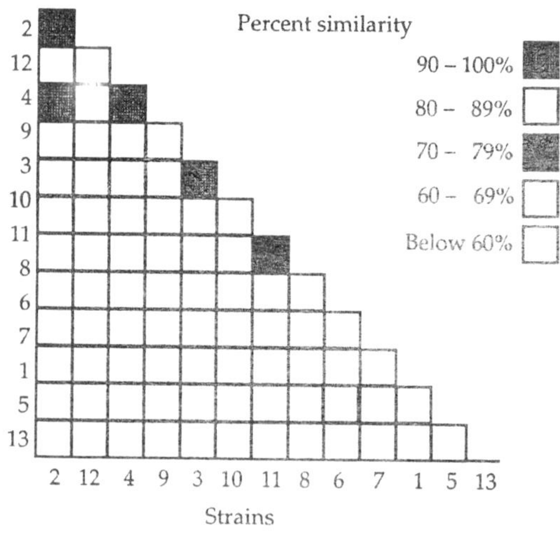

Fig. 5.1 Thirteen different strains of bacteria are arranged according to their similarity indices. Each intersecting square is then shaded to show the degree of similarity existing between each pair. With this type of matrix, one can readily see clusters of similarity such as shown here among strains 2, 12, 4 and 9 and among strains 3, 10, 11, 8, 6 and 7.

GENETIC CLASSIFICATIONS

DNA Homology

In recent years, the phylogenetic relationships occurring in bacteria, even among those not sharing many biochemical properties, has been estimated by determing the degree of similarity existing between their nucleic acid sequences. In the most direct test for DNA homology (**Fig. 5.2**) the DNA from

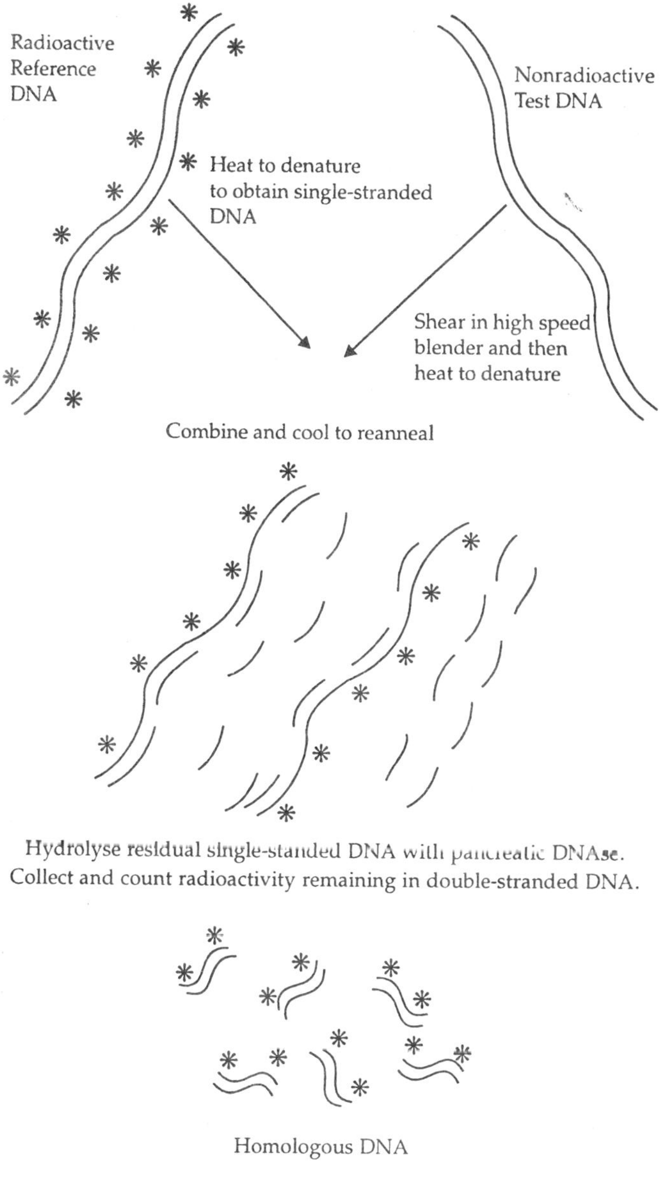

Radioactive Reference DNA

Nonradioactive Test DNA

Heat to denature to obtain single-stranded DNA

Shear in high speed blender and then heat to denature

Combine and cool to reanneal

Hydrolyse residual single-standed DNA with pancreatic DNAse. Collect and count radioactivity remaining in double-stranded DNA.

Homologous DNA

Fig: 5.2 The relatedness of two organisms can be determined by measuring the degree of DNA homology existing between the organisms. The reference organism is grown in a medium containing *tritiated thymidine* so that all of its DNA will be radioactively labeled with tritium. The test organism is grown in a medium containing no radioactively labeled components. The DNA is isolated from each organism and treated as described. Only those fragments of DNA from the test organism passessing sequences complementary to the reference DNA will reanneal to form double-stranded DNA.

one organism is made radioactive and is mixed with an excess of small DNA fragments obtained from the test organism. The mixed samples are heated to separate the strands of DNA and are then slowly cooled to allow the reformation of double-stranded regions by any homologous portions of the two DNAs. The degree of DNA homology between the two organisms is revealed by the rate and extent to which the test DNA will form stable double-stranded duplexes with that from the reference organism. There are several ways to determine the amount of radioactivity present in the double-stranded DNA species such as the use of chromatographic procedures that separate single stranded DNA from the double-stranded species, or the preferential degradation of the single-stranded molecules by the action of the S1 nuclease enzyme.

Ribosomal RNA

All bacteria possess ribosomal RNAs, whose structures and functions have been strongly conserved, that is, the sequences have changed slowly since the time any two organisms diverged from a common progenitor. Thus, the degree of identity of the rRNA sequences can be related to the time that the evolutionary tracks of two organisms separated. The relatedness of rRNAs (primarily 5S or 16S) can be easily estimated from the number of identical oligonucleotide spots seen following enzymatic digestion and separation of an rRNA. Rapid techniques for the determination of DNA sequence make feasible comparisons of the entire sequence for certain rRNAs. In this way, organisms can be compared for the number of base changes that have occurred in this portion of the genome, and this provides a more meaningful estimator of the genetic relatedness of two organisms.

Table—5.1 *Escherichia coli and Klebsiella Pneumoniae DNA Reactions with Erwinia Species*

DNA Reaction	*Relative Percent Binding, 60°C*
Escherichia colil/Escherichia coli	100
Escherichia colil/Erwinia Amylovora	27
Escherichia colil/Erwinia dissolvens	34
Escherichia colil/Erwinia aroiceae	16
Escherichia colil/Erwinia carotovora	17
Escherichia colil/Erwinia carnegieana	32
Escherichia colil/Salmonella Typhimurium	45
Escherichia colil/Proteus mirabilis	9
Klebsiella Pneumoniae/Klebsiella pneumoniae	100
Klebsiella pneumoniae/Erwinia dissolvens	34
Klebsiella pneumoniae/Erwinia carotovora	17

DNA Sequences

Certainly, determination of DNA sequence is not the type of technique that will find immediate application in the clinical laboratory. However, using cloning techniques it is possible to obtain portions of a genome that are specific for a given genus, species, or subspecies. Such cloned DNA fragments could provide the basis for simple and rapid tests for the presence of that type of organism in a clinical specimen. The ease, specificity and sensitivity of hybridization techniques to detect homologous sequences promise to make this type of approach practical even for routine use. It should also be possible to use the DNA from a drug-resistance determinant to test for the presence of that determinant in a clinical sample and, perhaps within the next decade, it will be possible to identify an organism and determine its drug-resistance properties without even having to wait for the organism to grow.

MAJOR GROUPS OF BACTERIA

This section presents of somewhat arbitrary listing of the major groups of bacteria of medical importance. It must be recognized that these bacteria represent a rather small part of the bacterial world and exclude myriad organisms of fascinating shapes and metabolic capabilities. The classification system used here is based on major morphological and biochemical properties and all of the organisms mentioned here are described in much more detailed manner in subsequent chapters. The purpose of this section is only to introduce the reader to the names and general properties of the major pathogens.

Mycoplasmas

The mycoplasmas (formerly called pleuro pneumonia-like organisms, or PPLOs) lack cell walls, and, as a result, the cells lack any defined shape. In addition, they are very small cells that can pass through many filters that retain other bacteria, a property resulting in their initial classification among the viruses. Similar to animal cells, mycoplasmas are surrounded only by a plasma membrane containing sterols—the only bacterial group containing such compounds in their cell membranes. As seen by rRNA homologies, most of the mycoplasmas are related to the gram-positive organisms that gave rise to *Bacillus* and *Lactobacillus*. This is a small group of organisms, which includes free-living forms as well as major animal pathogens in the genus *Mycoplasma*.

Actinomycetes

These gram-positive bacteria exhibit a branching, filamentous growth similar to that of the eucaryotic fungi. Pathogens among the lower actinomycetes include the generally anaerobic. *Actinomyces* and the aerobic, often acid-fast *Nocardia*. These form fairly extensive mycelial structures, but they tend to fragment into irregular shapes in older cultures. The higher actinomycetes include *Streptomyces* and *Micromonospora*, which remain mycelial and reproduce by externally borne asexual spores. These normally are soil inhibitants and generally are nonpathogenic. Indeed, the actinomycetes are the source of most of the antibiotics active against other bacteria. Also included in this grouping are the *Mycobacterium* whose extent of branching is not so prominent as in the higher actinomycetes.

Mycobacteria :

The mycelial character of Mycobacterium is usually not as obvious as that of the higher actinomycetes because the very limited mycelium breaks up into unicellular forms. A prominent characteristic of this group is the large amount (up to 40% of the weight of the cell) of complex waxy lipid covalently associated with the cell surface carbohydrate. This renders the cells difficult to stain or destain and is responsible for the property of acid-fastness. Because of their hydrophobicity, these organisms tend to grow in clumps as pellicles on the surface of the medium (an apparent advantage for these obligately aerobic organisms). Included in the genus *Mycobacterium* are the agents of tuberculosis and leprosy.

Spirochetes

The spirochetes are long, thin, flexible helical organisms, shaped like a corkscrew. Their shape is maintained in part by an axial filament, which is wrapped around their cell body. This filament, lying between the cell membrane and the cell wall, is actually a bundle of flagella that originates from the poles of the cell. Many spirochetes are too thin to be readily visible in stained preparations, but can be seen by dark-field microscopy. This diverse group of organisms includes major pathogens such as *Treponema, Borrelia,* and *Leptospira*.

Eubacteria

The eubacteria is the largest group and comprises bacteria possessing a rigid cell wall and which, if motile, employ flagella as the organ of motility. They include both gram-positive and negative cells and may be either rods or cocci. Even if we ignore the stalked and budding forms, which are not pathogens, the eubacteria still represent a very diverse group.

Pseudomonads :

Pseudomonads are gram-negative rods that are motile by means of polar flagella. These bacteria have only an aerobic form of metabolism and are characterized by their ability to digest an extremely broad range of substances. They are able to survive and even to grow in some very unusual environments. Such as jet fuel and harsh detergents. All of these organisms are free-living, usually in water, but some members of the genus Pseudomonas produce serious infections in impaired patients, such as those with burns or cystic fibrosis.

Vibrios :

Vibrios are similar to pseudomonads, but are curved or comma-shaped. One important pathogen is *Vibrio cholerae,* the cause of Asiatic cholera.

Spirilla :

Members of the genus *Spirillum* are S-shaped and are not to be confused with the spiral spirochetes. This group of organisms is similar to pseudomonads except in their shape. Only *Spirillum minus,* the cause of rat bite fever, is known to produce human disease.

Enteric Bacteria :

The enteric bacteria include a large number of related organisms, many of whom normally reside in intestinal tracts. They are gram-negative, facultatively anaerobic rods. This term means that they can switch their metabolic pathways in response to the presence or absence of oxygen. Most are motile with a peritrichous distribution of flagella, which means that the flagella are distributed all over the cell, rather than only at their poles. The group includes the general *Escherichia, Salmonella, Shigella, Enterobacter, Klebsiella,* and *Proteus,* as well as a number of intermediate forms that arise from the rather promiscuous genetic exchange among these organisms. All of these organisms are potential pathogens.

Lactic Acid Bacteria :

The group of biochemically related bacteria comprises both gram-positive cocci and short rods. All ferment sugar with the production of lactric acid although some form additional acids and CO_2. These organisms lack the enzyme catalase and thus prefer to grow at low levels of oxygen where their production of toxic H_2O_2 is greatly reduced. Members of the genus *Streptococcus* include many pathogenic species as well as others that are members of the normal human flora. Members of the

Lactobacillus genus are often employed in the preparation of foods such as sauerkraut, pickles, and olives, and in the fermentation of soy bean carbohydrates to form soy bean carbohydrates to form soy sauce.

Members of the genus *Listeria* are unrelated metabolically to the lactic acid bacteria but because this genus is comprised of non-spore-forming, gram-positive rods, Bergey's Manual groups them in the same section. *Listeria monocytogenes* may cause many types of human infections and it is recognized as the most common cause of bacterial meningitis occurring in immuno-suppressed renal transplant patients.

Micrococci :

These gram-positive cocci generally are aerobic organisms and, unlike the streptococci, possess catalase and a cytochrome system. The major pathogen in this group is *Staphylococcus aureus*.

Endospore-Forming Bacteria :

At least five genera of bacteria can produce endospores, but only two are of medical importance. These gram-positive rods are the aerobic genus *Bacillus* and the anaerobic genus *Clostridium*. Endospores are not an obligate part of the cell's reproductive cycle (as is the case in actinomycetes, fungi, and higher plants). Instead, the spores are a survival mechanism formed in response to nutrient deprivation. The spores usually are extremely resistant to heat, chemicals, and radiation. Since many of the organisms that produce endospores also produce very potent toxins, food preservation requires sufficient treatment to ensure the destruction of all spores. Among the pathogens in this group are the causative agents of anthrax, botulism, tetanus, and gas gangrene.

Diphtheroids :

The diphtheroids are gram-positive rods or club-shaped organisms whose cell division occurs with a sort of snapping action, so that cells remain attached to one another but at unusual angles. This group includes *Corynebacterium diphtheriae* and many saprophytes common to the throat and skin.

Obligate Anaerobes :

This group comprises the most prevalent bacteria in the human body, consisting of the major inhabitants of both the oral cavity and the intestinal tract. These organisms are so exquisitely sensitive to the presence of oxygen that they cannot be obtained in culture except when specialized techniques are used for their collection and growth. Although part of the normal bacterial flora, they are serious, life-threatening pathogens if they become established in sites other than their normal niche. Infections caused by these organisms are accompanied by a very fetid odour. Major members of this group include members of *Bacteroides,* the most profuse organism in the intestinal tract, and *Flexibacterium* and *Capnobacterium* in the oral cavity. Most of these organisms are gram-negative, and their metabolism invariably is fermentative.

Gram-Negative Cocci :

The primary pathogens found in this group of bean-shaped cocci are *Neisseria meningitidis,* the cause of epidemic meningitis, and *Neisseria gonorrhoeae,* the etiologic agent of the venereal disease, gonorrhea. There are, however, a number of saphrophytic members of the genera *Neisseria* and *Branhamella,* which are found as normal flora in the human body, particularly the oral cavity. Such organisms cause disease only in immuno-deficient persons.

Small, Immotile Gram-Negative Rods :

These organisms are distrinct from the enteric bacteria in their metabolic properties, their smaller size, and their extensive nutritional requirements. Important pathogens in this group include members of the genera *Brucella, Yersinia, Pasteurella, Haemophilus,* and *Bordetella.*

Obligate Intracellular Parasites :

Members of the genera *Rickettsia* and *Chlamydia* are important pathogens, which are able to divide only when growing within a eucaryotic cell. Their requirement for a host cell appears to stem from their need to obtain ATP or other complex nutrients or biosynthetic intermediates from the host.

6

Morphology of Bacteria

Bacteria belong to the class of organisms known as the *Schizomycetes (schizo,* fission, and *mycetes,* fungi). The organisms are single-celled and reproduce normally by transverse or binary fission.

. The class Schizomycetes is divided into ten orders. The largest order is the *Eubacteriales;* it includes most of the common bacterial species.

Bacteria are typically unicellular plants, the cells being usually small, sometimes ultramicroscopic. They are frequently motile. By means of modern techniques, a true nucleus has been demonstrated in bacterial cells. Individual cells may be spherical or straight, curved or spiral rods. Cells may occur in regular or irregular masses, or even in cysts. Where they remain attached to each other after cell division, they may form chains or even definite trichomes. The latter may show some differentiation into holdfast cells and into motile or non-motile reproductive cells. Some grow as branching mycelial threads whose diameter is not greater than that of ordinary bacterial cells, i.e., about 1μ. Some species produce pigments. The true purple and green bacteria possess photosynthetic pigments much like or related to the true chlorophylls of higher plants. The phycocyanin found in blue-green algae does not occur in the Schizo mycetes. Multiplication is typically by cell division. Endospores are formed by some species of Eubacteriales. Sporocysts are found in Myxobacteriales. Bacteria are free-living, saprophytic, parasitic, or even pathogenic. The latter types cause diseases of either plants or animals.

Filament Formation : Cells that reproduce and divide in a normal manner may be induced to grow in filaments by changing the conditions of the medium. According to Webb (1953)

> ... the division of the bacterial cell follows a complex sequence, which in many respects, resembles that occurring in the cellular reproduction of higher forms. It is now known, for example, that bacterial cell division entails division of the nuclear element, division of the cytoplasm, secretion of new cell wall material, and the separation of the daughter cells.
>
> Some or all of the events of this sequence are readily thrown out of balance, or even completely inhibited. Thus bacteria, particularly the rod-shaped organism, may be induced to elongate into filaments by various treatments which apparently inhibit cell division but which do not inhibit growth. Such an effect is produced by various chemical substances, by sub-bacteriostatic concentration of certain antibacterial agents, as, for example, methyl violet, sulfonamides, *m*-cresol, penicillin, irradiation, ad higher temperatures of incubation.
>
> These changes in morphology induced by chemical substances are usually temporary, since reversion to normal form occurs promptly when the filamentous bacteria are subcultured in the absence of the inhibitory agents. Irradiation, on the other hand, may give rise to a temporary or permanent induction of filamentous cells.
>
> From observations such as these the concept has arisen that bacterial growth, in the sense of an irreversible increase in cell substance or volume, and cell division may be considered to some extent as separate and independent processes; at least, in so far as growth may occur either with or without the operation of the cell division mechanism

Variation in the magnesium (Mg) content of the medium may exert a marked effect on cell division of some bacteria. In a Mg-deficient medium, Gram-positive rods grow in the form of long filaments. Such filaments revert to normal forms when transferred to the same medium supplemented with suitable concentrations of Mg. Filament formation is enhanced by the addition of zinc and cobalt. Inhibition of cell division occurs also in media supplemented with an excess of Mg.

Deibel et al. (1956) produced filamentous *Lactobacillus leichmannii* in the absence of vitamin B_{12}. Reversion to the normal cell form occurred on the addition of either vitamin B_{12} to a medium lacking the growth factor or of an excess of the desoxyriboside thymidine.

Shape of Bacteria : Bacteria exhibit three fundamental shapes: (1) spherical, (2) rod, and (3) spiral or curved rod. All bacteria exhibit pleomorphism in more or less degree under normal or other conditions, but a bacterial species is still generally associated with a definite cell form when grown on a standard medium under controlled conditions.

The spherical bacteria (singular, coccus; plural, cocci) divide in one, two, or three planes, producing pairs or chains, clusters, or packets of cells. Some are apparently perfect spheres, others are slightly elongated or ellipsoidal in shape.

The streptococci divide in only one plane. They grow normally in pairs or chains. Depending upon the species, the distal ends of each pair may be lancet-shaped, or flattened at the adjacent sides to resemble a coffee bean.

The staphylococci divide in two planes, producing pairs, tetrads, or clusters of bacteria, the latter resembling bunches of grapes.

The sarcinae divide in three planes, producing regular packets These are cubicle masses with one layer of bacteria atop another.

The rod forms also show considerable variation. A rod is usually considered to be a cylinder with the ends more or less rounded. Some rod forms are definitely ellipsoidal in shape. The ends of rods also show considerable variation. Some species are markedly rounded; others exhibit flat ends perpendicular to the sides. Gradations between these two forms may be seen.

Rods may show marked variation in their length/width ratio. Some rods are very long in comparison to their width; others are so short they may be confused with the spherical forms.

The shape of an organism may also vary depending upon certain environmental factors, such as temperature of incubation, age of the culture, concentration of the substrate, and composition of the medium. Bacteria usually exhibit their characteristic morphology in young cultures and on media possessing favourable conditions for growth.

Young cells are, in general, larger than old organisms of the same species. As a culture ages, the cells become progressively larger until a maximum is reached, after which the reverse effect occurs. Bacterial variations resulting from changes in age are only temporary; the original forms reappear when the organisms are transferred to fresh medium.

Size of Bacteria : Bacteria vary greatly in size according to the species. Some are so small they approach the limit of visibility when viewed with the light microscope. Others are so large they are almost visible with the normal eye. However, the sizes of the majority of bacteria occupy a range intermediate between these two extremes. Regardless of size, none can be clearly seen without the aid of a microscope.

A spherical form is measured by its diameter; a rod or spiral form by its length and width. Calculation of the length of a spiral organism by this method gives only the apparent length, not the true length. The true length may be computed by actually measuring the length of each turn of the spiral. Mathematical expressions have been formulated for making such computations.

The method employed for fixing and staining bacteria may make a difference in their size. The bacterial cell shrinks considerably during drying and fixing. This will vary somewhat depending upon the type of medium employed for their cultivation. Shrinkage generally averages about one-third of the length of the cell as compared to an unstained hanging-drop preparation. Young cells of *bacillus megaterium* may shrink from 15 to 25 per cent when transferred from nutrient broth to the same medium containing sodium chloride in 2 *M* concentration.

Measurements show some variation depending upon the staining solution used and the method of application. In dried and fixed smears, the cell wall and slime layer do not stain with weakly staining dyes such as methylene blue but do stain with the intensely staining pararosaniline, new fuchsin, crystal violet, and methyl violet. The great majority of bacteria have been measured in fixed and stained preparations. In some instances dried, negatively stained smears have been used. Therefore, the method employed should be specified when measurements of bacteria are reported; otherwise the results will be of doubtful value.

The unit for measuring bacteria is the micron. It is expressed by the symbol μ. It is 0.001 mm. or 0.0001 cm., A millimicron is 0.001 μ or 0.000001 mm. It is expressed by the symbol mμ.

Some bacteria measure as large as 80 μ in length; others as small as 0.2 μ. However, the majority of the commonly encountered bacteria, including the disease producers, measure about 0.5 μ in diameter for the spherical cells and 0.5 by 2 to 3 μ for the rod forms. Bacteria producing spores are generally lager than the non-spore-producing species. The sizes of some common species in dried and stained smears are as follows: *Escherichia coli*, 0.5 by 1 to 3 μ: *Proteus vulgaris*, 0.5 to 1 by 1 to 3 μ; *Salmonella typhosa*, 0.6 to 0.7 by 2 to 3 μ: *Streptococcus lactis*, 0.5 to 1 μ in diameter; *S. Pyogenes*, 0.6 to 1 μ in diameter; *Staphylococcus aureus*, 0.8 to 1 μ in diameter; *Lactobacillus acidophilus*, 0.6 to 0.9 by 1.5 to 6 μ; *Bacillus subtilis* rods, 0.7 to 0.8 by 2 to 3 μ, spores, 0.6 to 0.9 by 1 to 1.5 μ; *B. megaterium* rods, 0.9 to 2.2 by 1 to 5 μ, spores, 1 to 1.2 by 1.5 to 2 μ; *B. anthracis* rods 1 to 1.3 by 3 to 10 μ, spores, 0.8 to 1 by 1.3 to 1.5 μ.

The most commonly employed method for measuring bacteria is by means of an ocular micrometer. Measurements may also be made by using a camera-lucida attachment and drawing oculars, or by projecting the real image on a screen and measuring the bacteria.

The same factors that cause variations in the shape of bacteria also affect their size. With few exceptions, young cells are much larger than old or mature forms. Cells of *B. Subtilis* from a 4-hr. culture measure five to seven times longer than cells from a 24-hr. culture. Variations in width are less pronounced.

The organism *Corynebacterium diphtheriae* is a notable exception to the rule of decreasing cell size with age.

Variations in cell size with age are due to a variety of factors. The major causes appear to be changes in the environment with the accumulation of waste products. An increase in the osmotic pressure of the medium will also cause a decrease in cell size and may very well be the most important factor.

The Bacterial Cell

Bacteria do not show the same morphological picture. Differences in structure exist between species. It is generally agreed that a bacterial cell consists of a compound membrane enclosing cytoplasm and nuclear material and often containing various granules, fat globules, and one or more vacuoles. In addition, some species contain resistant bodies known as spores, and some have one or more organs of locomotion called flagella.

The term protoplasm is used to indicate the thick viscous semifluid or almost jelly-like colourless, transparent material which makes up the essential substance of both the cell body and the nucleus, including the cytoplasmic membrane but not the cell wall. It contains a high percentage of water and holds fine granules in suspension.

Cytoplasmic Membrane : This membrane appears in young cells as an interfacial fluid film, becoming thicker and denser as surface-active material accumulates. It is finally converted into a firm structure. The membrane is believed to be composed mainly of lipide and protein. Polysaccharide has not been demonstrated as a component.

The membrane is acid in reaction because of its content of ribonucleic acid. It stains deeply with basic and neutral dyes over a wide range of pH. The membrane stains Gram-positive in Gram-positive bacteria and acid-fast in acid-fast organisms. It is a semipermeable membrane and is principally responsible for the Gram and acid-fast reactions. When a cell is plasmolyzed by immersion in a hypertonic solution, this membrane is drawn is with the cytoplasmic constituents. The thickness of the membrane varies even in a single cell. Measurements on a strain of *Bacillus cereus* at various stages of development ranged from 5 to 10 mμ in thickness.

Cell Wall : The cell wall is a more rigid structure and is responsible for the form of the bacterial body. It behaves as a selectively permeable membrane and apparently plays a fundamental role in the life activities of the cell.

The cell wall has a low affinity for dyes, which means that it is probably not stained in some of the usual staining procedures. It is lightly stained by certain basic dyes such as basic fuchsin and the methyl violets. Where deep staining of the wall is desired, the use of a mordant, such as tannic acid, is necessary. The mordant not only increases the affinity of the cell for dye, but it may increase the thickness of the wall.

The cell wall accounts for an average of about 20 per cent of the dry weight of bacteria and represents the major structural component. In thickness, it ranges from 10 to 23 mμ, depending upon the species.

According to Salton (1952, 1953), chemical analyses of cell walls have revealed differences in Gram-positive and Gram-negative bacteria. Cell walls of Gram-positive bacteria are lacking in aromatic and certain sulfur-containing amino acids, arginine, and proline. On the other hand, cell walls of Gram-negative bacteria show the presence of aromatic and sulfur-containing amino acids, arginine, and proline.

Gram-negative cell walls are generally richer in lipides than Gram positive bacteria.

Cell walls of a number of Gram-positive and negative bacteria contain the amino acid diaminopimelic acid.

Polysaccharides have been detected in both Gram-positive and Gram negative bacteria. The polysaccharide is determined as reducing substances after acid hydrolysis. Some polysaccharides yielded only one reducing sugar; others yielded two or more sugars.

Gram-positive organisms gave rhamnose, galactose, and glucose; glucose only; rhamnose only; arabinose, galactose, and mannose. Gram-negative bacteria yielded galactose and glucose; galactose, glucose, mannose, and rhamnose.

In addition, all organisms studied contained an amino sugar or hexosamine. Generally, the walls of Gram-positive bacteria are richer in hexosamine than the Gram-negative forms.

Work (1957) pointed to the existence in Gram-positive bacteria of a common basal structure containing the following constituents: a hexosamine component comprising glucosamine and muramic acid and sometimes also galactosamine; a peptide component made up of alanine, glutamic acid, and either diaminopimelic acid or lysine with sometimes also glycine, aspartic acid, or serine; and usually a polysaccharide containing not more than four different sugar residues. Other substances may also be attached to the walls, as for example the protein antigens.

Smithies, Gibbons and Bayley (1955) reported a relatively high nitrogen content in the walls of several halophilic bacteria which indicated that the cell material was predominantly protein. They contained only small amounts of lipides. The cell walls were lipoprotein.

Barkulis and Jones (1957) found that approximately one-third of streptococcal cell walls was made up of rhamnose and hexosamine. The remaining two-thirds was protein in nature.

Capsules : Extracellular material of a slimy or gelatinous nature is formed by many bacteria, especially those producing mucoid growths. This material may remain firmly adherent as a discrete covering layer on each cell, or it may part freely from the cells. In the former case it is known as a capsule; in the latter, as free slime or gum.

Capsule and slime are believed to be distinct from the morphological and biochemical point of view. The capsule is a part of the cell, the slime a secretion. According to Klieneberger-Nobel (1948), capsules are of definite shape, of more or less definite density throughout, and of definite outline, where as slime envelopes are amorphous and can be drawn out into manifold structures, are most concentrated in the vicinity of the bacterial cells, and decrease in density with increasing distance from the cell.

Broth cultures of capsule-producing organisms are usually stringy in texture, and agar colonies exhibit a very moist, glistening surface which is described as mucoid. Capsule formation is dependent upon the composition of the medium but especially the variant phase of the organism. Some disease-producing organisms form large capsules in culture media rich in animal fluids. Others produce prominent capsules when cultures are incubated at low temperatures (4 to 20 °C).

Chemical analyses of capsular material from a number of bacteria show wide differences in composition. For this reason it is impossible to make statements which apply to all bacteria. In some organisms the capsular material appears to be a glycoprotein; in others, a protein-polysaccharide complex; in still others, a polysaccharide framework with the spaces filled in by a larger amount of glutamyl polypeptide.

Capsular material is difficult to distinguish from those gums which flow away from the cells as thy are formed. Organisms producing gums do so when grown in sugar solutions. Some organisms produce gums only in the presence of a specific sugar; others produce gums in the presence of any one of several sugars. In the absence of sugar, usually very little, if any, gum is formed. Organisms producing gums of this type are the cause of considerable losses in the sugar industry. The increased viscosity produced by the gum interferes with the filtration of the sugar solution.

The species commonly encountered in sugar-cane juice is *Leuconostoc mesenteroides*. The cells are surrounded by a thick, gelatinous, colourless polysaccharide consisting of dextran (glucose polymer).

The formation of gums is of common occurrence by soil bacteria. From 5 to 16 per cent of such forms have been shown to be capable of synthesizing gums from sugars.

Bacterial protoplasts : When the cell wall is damaged, the protoplasm usually disintegrates. However, methods are available for removing the cell membranes without destroying the vital nature of protoplasm. The term protoplast is used to indicate living protoplasm exclusive of the cell membranes.

Action of Lysozyme : Some bacteria are rapidly lysed or dissolved by the action of lysozyme. Weibull (1953) reported that lysozyme possessed a specific depolymerizing action on the cell wall and that this appeared to be the only portion of the cell that was affected by such treatment. As a result of the destruction of the wall, the protoplasts were liberated.

Protoplasmic structures are not very stable. Consequently, if protective agents were not employed, destruction of the walls was accompanied by a rapid lysis of the protoplasts, followed by the liberation of most of the cell protein and nucleic acid in soluble form., This could be prevented by employment of the enzyme in a 0.2 *M* solution of sucrose or cane sugar. After digestion of the cell walls, the living protoplasts rounded up into spheres.

Spiegelman, Aronson, and Fitz-James (1958) found that digestion of protoplasts of *Bacillus megaterium* led to the liberation of nuclear bodies of the protoplasts. Such bodies were collected by centrifugation for 5 min. at 10,000 × g.

Properties of Protoplasts : The difficulty in handling and studying protoplasts is their extreme fragility and sensitivity to osmotic shock, shaking, centrifugation, and aeration, Removal of the cell wall does not change the structure and capabilities of the protoplasm. Permeability, respiration, and spore formation appear to be the same for protoplasts and intact cells. Also both can support the development of bacteriophages. Under special conditions the protoplasts grow and probably divide like intact cells. However, there is no evidence that protoplasts form colonies. The metabolism of protoplasts and intact cells appear to be very similar but probably not identical.

Filterability of Protoplasts : Sinkovics (1958) reported the spontaneous occurrence of units in aged cultures of *Escherichia coli* which conformed to the description of artificially induced bacterial protoplasts. The disintegration of cells from aged cultures was preceded by swelling of the cell and rupture of the rigid cell wall. Centrifugation of the culture gave a supernate which, after filtration, contained units capable of regeneration when placed in fresh medium. The smallest unit capable of regeneration measured about 350 mμ in diameter. The units underwent fusion before cell-wall formation occurred.

The results supported the assumption that aged *E. coli* cultures could survive in the form of unit having no cell walls and which, under adequate conditions, regenerated into vegetative forms.

Polysaccharide Structures : Polysaccharides occur (1) in cell wills, (2) extracellularly in capsules and gums, and (3) inside of bacterial cells. The first two have already been discussed.

Pennington (1949) revealed the presence of polysaccharides by treating bacteria with sodium metaperiodate followed by staining with sulfite decolourized basic fuchsin. In *bacillus cereus* the polysaccharide was concentrated in the cytoplamic membrane as well a in the cell wall.

Selective staining of polysaccharide in the cell is said to depend upon the oxidizing action of periodate on such chemical configurations as α, β glycol and α-hydroxyketones. Polyaldehydes generated by this selective oxidation react with sulfite-decolourized fuchsin. Polysaccharide areas in the cell are coloured red by the stain.

Nucleus : The question of the presence of a well-defined nucleus in bacteria has been the subject of investigations by bacteriologists almost from the beginning of bacteriology.

Some of the earlier cytologist maintained that bacteria were very primitive organisms devoid of nuclei and consisting simply of cytoplasm, granules, and vacuoles. This view was based on their failure to observe a nucleus in a bacterial cell. Others held the view that the nuclear material was present in a diffuse form throughout the cytoplasm. Still others believed that the whole cell should be regarded as a "naked nucleus", corresponding to the nucleus of higher organisms. The naked nucleus is regarded as a primitive form of living matter. Since bacteria have the structural and physio-logical attributes of true cells, this concept cannot apply to these organisms.

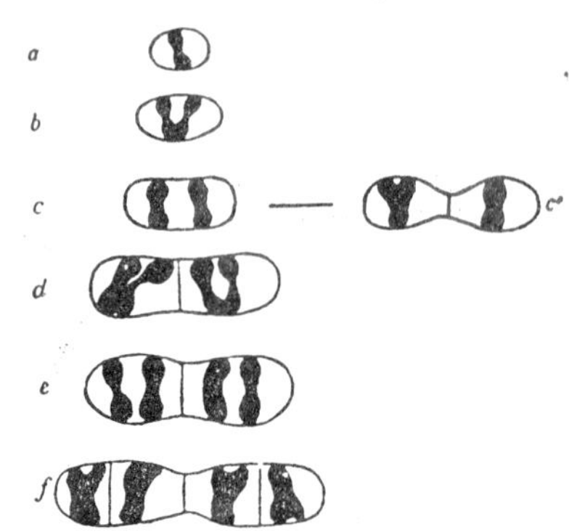

Fig. 6.1 *Escherichia coli*. Diagram of successive division stages of the chromatinic bodies from the beginning of the lag phase, after transfer to a fresh nutrient medium, to the first division of the growing organism. c–c′ and c–f are alternative modes of development, c–f being that most commonly followed.

Much of the confusion was caused by the inadequacy of the staining procedures. By means of the HCl-Giemsa staining technique of Piekarski, many obser-vations have been reported demonstrations the presence of chromatinic structures in bacteria.

Robinow (1944) prepared wet smears of *Escherichia coli*. Slides were fixed in osmic acid vapour, dried, and immersed in normal HCl for about 9 min. at 53 to 55° C., then washed and stained in 1:20 Giemsa solution for 10 to 60 min., depending on the staining properties of the specimen.

The chromatinic structures in *E. coli* from old cultures were too small to be resolved accurately. After transfer to fresh medium the chromatinic structures increased in size and gave rise to short, often dumbbell-shaped rods or chromosomes, which multiplied by splitting lengthwise in a plane more or less parallel to the short axis of the cell. A single cell of *E. coli* contained one chromatinic body or one or two pairs of these representing primary and secondary division products.

The Smith (1950) technique consisted of fixing the smear in osmium tetroxide vapour, immersion in HCl, mordanting in dilute formaldehyde, and staining with aqueous basic fuchsin. The method was said to possess certain advantages over the procedure of Robinow.

Another cause of confusion in the recognition of nuclear structures in bacteria was the lack of appreciation for the ages of the cultures. At certain times nuclear structures cannot be seen. In general, most of the early observations of tidy, intelligible "nuclei" were made on preparations from very young cultures, whereas haphazardly scattered, unintelligible granules of chromatin were persistently observed in preparations made from cultures of the same bacteria beyond the logarithmic growth phase.

Recognition of the fact that the configuration of nuclear material in bacteria might change with age removed one of the chief causes of confusion.

Direct division of the chromatin bodies of *E. Coli.* has been demonstrated by Mason and Powelson (1956) in a series of remarkable phase contrast photomicrographs. These observations were made on living bacteria. The nuclear areas in the dividing cells appeared to be as clearly defined as the areas in fixed, hydrolyzed, and stained cells.

Vacuoles : Vacuoles have been identified in young bacteria. They are cavities in the protoplasm and contain a fluid known as cell sap. As the cells approach maturity, some of the water-soluble reserve food materials manufactured by the cell dissolve in the vacuoles. Insoluble constituents precipitate out as cytoplasmic inclusion bodies.

Metachromatic Granules : The best-known inclusion bodies in bacterial cells are known as volutin or metachromatic granules. The granules are small in young cells and become larger with the age of the culture. They are believed to originate in the cytoplasm of young cells and to localize in the vacuoles of mature forms. The granules show a strong affinity for basic dyes, indicating that they are acid in character. They are usually considered to b a reserve source of food.

Grula and Hartsell (1954) found the granules to be composed of metaphosphate or another form of inorganic phosphate, some fat, and possibly small amounts of protein. Their presence and size in cells were related to the phosphate concentration of the growth medium in the presence of an energy source and specific divalent ions (Mn and Zn). Older cells possessed larger granules. Their basophilic nature did not depend on either ribonucleic acid (RNA) or desoxyribonucleic acid.

On the other hand, Widra (1959) found metachromatic granules to contain protein-bound lipide, RNA, and polyphosphates.

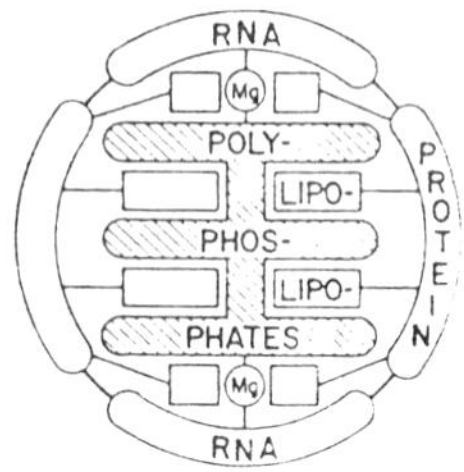

Fig.6.2 Composite representation of a metachromatic granule.

Fat Globules. Bacteria are capable of storing fat in the form of globules. Fat globules may be demonstrated in 24-hr. Cultures and usually reach a maximum in about 48 hr.

Some cells may contain only one large globule; others may show the presence of a number of small, scattered globules.

It is generally believed that fat is stored as reserve food material. Globules usually cannot be demonstrated in young, vigorously growing cells. As cells age and slow down in activity, fat globules appear in the cytoplasm and may be recognized by appropriate staining.

Motility : Bacterial motion is generally associated with the presence of organs of locomotion known as flagella (singular, flagllum), They were first observed in stained preparations by Cohn (1875). The presence of flagella does not mean necessarily that the organisms are always motile, but it indicates a potential power to move.

Independent bacterial motion is a true movement of translation and must be distinguished from the quivering or back-and-forth motion exhibited by very small particles suspended in a liquid. This latter type of motion is called Brownian movement (after Brown, (1828) and is caused by the bombardment of the bacteria by the molecules of the suspending fluid.

Properties of Flagella : Flagella are very delicate organs and easily detached from the cell. In the stained condition they are long, slender, undulating organs. They are directed backward to the direction of motion at an angle of about 45°. Reversal of direction occurs by swinging the flagella through an angle of about 90°. Turning movements take place by swinging the flagella forward on one side only. They propel the organism by a spiral or corkscrew motion.

The thickness of flagella varies from species to species. In *Proteus vulgaris* they measure about 12 mμ. This figure is considerably below the shortest wave length of visible light and explains why flagella cannot be seen in hanging-drop preparations or in smears stained by the usual simple procedures. When special staining methods are employed, sufficient dye becomes deposited on the flagella to make their diameters greater than the wave length of visible light. They may then be seen under a light microscope.

Chemistry of Flagella : Flagella and bacterial bodies differ in composition. Flagella break up on boiling or when exposed to pH values below 4 or above 11. Their composition is largely protein, having a molecular weight of about 41,000. Weibull (1949) found the flagella of *P. vulgaris* to be composed of 98 per cent protein, traces of carbohydrate and fat, and no phosphorus. The protein contained only 14 known amino acids. It is an incomplete protein, lacking in some of the essential amino acids.

It is well established that the *H* antigens of bacteria are associated with the flagella and the *O* antigens with the bodies. Purified flagella are agglutinated by *H* antiserum but not by *O* antiserum; *O* antigens are not agglutinated by *H* antiserum. This is another indication that flagella and bacterial bodies differ in composition.

Origin of Flagella : Some believe flagella originate from the cell wall; others believe they traverse the cell wall into the protoplasm.

Flagella differ chemically both from the cell wall and the protoplasm. Two observations have been made relative to the site of origin of flagella.

Electron micrographs by van Iterson and others show the flagella penetrating the faint outer zones and extending into the cytoplasm. If the outer zone is the cell wall, then the flagella have their origin in the cytoplasm.

Weibull (1953) showed that removal of the cell wall of some bacteria by means of lysozyme produces a spherical protoplast which still retains the flagella of the treated cell. The obvious conclusion is that flagella have their origin in some cell structure deeper than the cell wall.

Number and Arrangement of Flagella : The number and arrangement of flagella vary with different bacteria, but they are generally constant for each species. Some have only one flagellum; others have two or more flagella.

In rod-shaped cells the flagella arise either at one or both poles, or are distributed laterally with the poles being generally bare. In some species flagella are located both laterally and at the poles. A species may show considerable variation in the number and arrangement of flagella. Single *Alcaligenes* cultures may contain forms with a polar flagellum only some with several lateral flagella and some with both lateral and polar flagella. Sometimes a species may show cells which are flagellated in one environment and nonflagellated in another.

Leifson, Carhart, and Fulton (1955) reported the presence of four definite types of curvature in the flagella of *proteus vulgaris*. Individual organisms may have more than one type of flagella, and individual flagella may have one or two types of curves.

Environmental factors, particularly pH, may change the curvature of the flagella on some strains, but not on all strains. In acid media the curly curvature tends to predominate; in alkaline media the normal predominates.

Organisms have been classified on the basis of the number and arrangement of flagella as follows:

Monotrichous—a single flagellum at one end of the cell.
Lophotrichous—two or more flagella at one end or both ends of the cell.
Amphitrichous—one flagellum at each end.
Peritrichous—flagella surrounding the cell.

Staining of Flagella : The staining of flagella is a difficult technique, especially in the hands of the beginner. For this reason many methods have been proposed. Regardless of the method employed, the film must first be treated with a mordant to make the flagella take the stain heavily. Mordants consist usually of a mixture of tannic acid and some metallic salt. In some methods the mordant and stain are applied separately; in others they are combined in one solution.

The Pijper Theory of Motility : In a series of investigations Pijper et at. (1957) questioned the belief that flagella are responsible for motility. He added methyl cellulose to a culture of a motile organism to increase the viscosity of the medium. This treatment decreased the motility of the cells. Under these conditions the cells exhibited a gyratory undulating movement like other aquatic creatures. He concluded that flagella were not organs of locomotion but only artifacts—useless appendages, polysaccharide twirls—the result, not the cause, of bacterial motility.

Notwithstanding the findings and conclusions of Pijper, evidence at present appears to be overwhelmingly in favour of flagella as organs of locomotion.

As has already ben stated, Weibull (1948, 1949) reported the flagella of *P. vulgaris* to be composed of 98 per cent protein. This contradicts Pijper's statement that flagella are formed from the carbohydrate slime layer that is peeled off into a number of thin, wavy threads.

Several investigators, among them Labaw and Mosley (1955), by means of electron microscopy, demonstrated the presence on *Brucella bronchiseptica* of uniform flagella having an external contour of a counter-clockwise or left-handed triple helix. The average periodicity along the length of the flagella was 19 mμ, with an average diameter of 13.9 mμ.

Motion of Colonies : Several organisms have been described which exhibit colonial motility when grown on a solid medium.

Shinn (1938) prepared lapse-time motion pictures of individual colonies of *Bacillus alvei* grown on agar plates and measured their velocities. The linear motions of colonies measuring 0.2 to 0.5 mm. in diameter averaged about 14 mm. per hr. Comparing this figure with the speed of individual cells of other species of motile bacteria gave the following results:

Salmonella typhosa	65 mm. per hr.
Bacillus megaterium	27 mm. per hr.
B. alvei (colonies)	14 mm. per hr.

The colonies exhibited not only linear motion but also a slow rotary movement. The direction of rotation of 200 to 300 colonies observed was counter clockwise, with the exception of two colonies in which it was clockwise.

Turner and Eales (1941*a*, *b*) reported that the rotation of an aerobe occurred very early during growth. The cells segregated in small groups and aligned them-selves concentrically around a common center to form disk-like'plaques one or a few cells thick. The rate of rotation was gréater in smaller groups. As multi-plication continued, successive layers were gradually built up in terrace fashion and the colony grew in height. The colonies then began to migrate. When a colony migrated, it left a peculiar "track" on the surface of the agar. A small number of cells were left behind, mostly at the edges of the track, which formed two parallel lines separated by the width of the moving colony.

Fig. 6.3 Sketch of convoluted track of a wandering colony, showing two series of clockwise spirals followed by a final counterclockwise spiral. The colony had increased considerably in size after coming to rest and showed curved radial markings indicatig rotation. The toral length of the track was about 2 cm.

Typical migrating colonies pursued curved or spiral paths which were often very elaborate and of relatively great length, even 2 or 3 cm. The direction of rotation was either clockwise or counter clockwise. After wandering of a variable distance, a colony approa-ched the center of its spiral path with rapidly shortening radius, ceased to migrate, began to rotate around its center, lost its elongated shape, and increased in size to several times the width of the track at the end of which it was formed.

For more information see Murray and Elder (1949).

Endospores : Endospores are bodies produced within the cells of a considerable number of bacterial species. They are more resistant to unfavourable environmental conditions, such as heat, cold, desiccation, osmosis, and chemicals, than the vegetative cells producing them. However, it is debatable if such extreme conditions actually occur in nature. For instance, the resistance of spores to high temperatures is a laboratory phenomenon and probably never occurs in a natural environment.

The bulk of evidence indicates the existence of a close relationship between spore formation and the exhaustion of nutrients essential for continued vegetative growth. Sporulation is a defense mechanism to protect the cell when the occasion arises.

Spore formation is limited almost entirely to two genera of rod-shaped bacteria: *Bacillus* (aerobic facultatively anaerobic), and *Clostridium* (anaerobic or aerotolerant). With one possible exception, the common spherical bacteria do not sporulate. Some spore-bearing species can be made to lose their ability to produce spores. When the ability to produce spores is once lost, it is seldom regained. Sporulation is not a process to increase bacterial numbers because a cell rarely produces more than one spore.

Morphology of Spores : Spores may be spherical, ellipsoidal, or cylindrical in shape. The position of the spore in a cell may be central, subterminal, or terminal. A fully grown spore may have a diameter greater than that of the vegetative cell. This causes a bulging of the cell. The resulting forms are known as clostridium if central, and plectridium if terminal. As a rule, each species has its own characteristic size, shape, and position of the spore, but this is subject to variation under different environmental conditions.

Franklin and Bradley (1957), by means of electron microscopy of carbon replicas, reported that the spores of a majority of species of *Bacillus* and *Clostridium* are readily distinguished by surface patterns. The surfaces may be smooth or ribbed, with the ribs usually longitudinal.

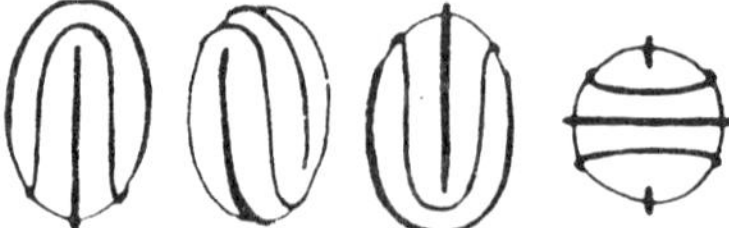

Fig. 6.4 Surface structure of a spore of *bacillus polymyxa*. From left to right: side view; same rotated a quarter turn from right to left; same rotated a further quarter turn; view of a pole.

Drawings of the surface sculpturing of spores of *B. polymyxa* are shown in fig. 6.4. the sculpturing consists of a single endless ridge in the form of two loops, similar to the marking on a tennis ball, together with two other separate ridges terminating within the loops. An electron micrograph of ultrathin sections of such spores, by van den Hooff and Aninga (1956), The spore coat consists of an outer and an inner layer separated by a space. The outer layer is sometimes called the exine and the inner layer the intine. The intine faintly follows the surface relief. The central core is separated from the intine by a regular nonosmophilic space. A peripheral spot may be observed in the core which probably represents nuclear material.

For more information see Bradley and Franklin (1958), Dondero and Holbert (1957), Hashimoto and Naylor (1958). Mayall and Robinow (1957).

Parasporal Bodies : When sporulatioon of *Bacillus laterosporus* is complete, the spores are cradled in canoe-shakped bodies.

Composition of Spores : Ross and Billing (1957), by means of refractive index measurements on spores and vegetative cells of *B. cereus, B. cereus* var. *mycoides*, and *B. megaterium*, found the values to be very high and comparable with that of dehydrated protein. This suggested that they contained much less water than the vegetative cells.

Strange and Dark (1956) demonstrated the presence of a hexosamine containing peptide in the spore coats of *B. megaterium* and *B. Subtilis*. The breakdown of an insoluble peptide complex might well be one of the first steps of the germination process. It was believed that the release of the hexosamine-amino acid complex was the result of the action of lysozyme present in the spores.

Enzymes of Spores : The presence of enzyme systems in *Bacillus* spores has been reported. Some of these are an inorganic pyrophosphatase that requires manganese for activation adenine ribosidase that hydrolyzes adenosine, an enzyme that functions possibly to lyse the sporangium and free the spore during germination, highly active alanine racemase that catalyzes the conversion of L-alanine to D-alanine, several glucose dehydrogenases, and an aldolase.

Sporulation process : Conditions necessary for sporulation in one species do not necessarily apply to another. The subject appears to be in such a state of confusion that it is impossible to discuss sporulation in terms of generalities.

The conditions which have been reported as favouring sporulation include addition of salts of metals such as manganese, chromium, nickel, etc., to the medium; shaking a culture of vegetative cells of sporing aerobes with distilled water at 37°C.; addition of tomato juice to a medium; incubating the cultures at an appropriate temperature; addition of calcium carbonate to a carbohydrate medium to prevent excessive accumulation of acid, and to maintain the pH at 5.5 or above; the necessity of oxygen; addition to the medium of certain amino acids; etc.

Germination of Spores : With the exception of some constituents such a high concentrations of calcium, dipicolinic acid, and in *Bacillus sphaericus,* α, ε-diaminopimelic acid, spores are similar to the vegetative cells in composition.

When a spore prepares itself for germination, it loses its refractility, which coincides with an imbibition of water. This stage is associated with a loss in heat resistance, stainability, and dry weight. Later the spore coat breaks, followed by the emergence from the spore case of a new germ cell which eventually matures into a vegetative cell.

Spore germination has been defined in various ways. According to Campbell (1957), "Spore germination may be regarded as the change from a heat resistant pore to a heat labile entity which may not necessarily be a true vegetative cell." Later development, leading eventually to the formation of a mature vegetative cell, is called outgrowth.

Some conditions which stimulate germination are as follows: (1) Treatment at 90 to 100°C. for 1 to 2 min. stimulates germination. (2) Spores which fail to germinate overcome this dormancy when activated by heat. (3) The use of certain agents such as alanine, glucose, and adenosine stimulates pore germination in most sporing species. In some species other amino acids may be substituted for the alanine. The same applies to glucose. (4) Yeast extract and mixtures of vitamin-free amino acids have also been shown to stimulate germination.

Spores germinate in a variety of ways. There is a considerable degree of constancy in the method of spore germination for each species. Lamanna (1940) classified the modes of germination as follows:

I. Spore germination by shedding of spore coat. Characteristics of this method are

   A. Spore does not expand greatly in volume previous to the germ cell breaking through the spore coat. The limit of volume increase of the spore may be considered to be twice its original volume.

   B. Spore coat does not lose all its refractive property previous to germination.

   C. After the second division of the germ cell, giving a chain of three organisms, the original spore coat, remaining attached to the cells, is visible for a long time after germination.

      1. Equatorial germination.
      2. Polar germination.
      3. Comma-shaped expansion.

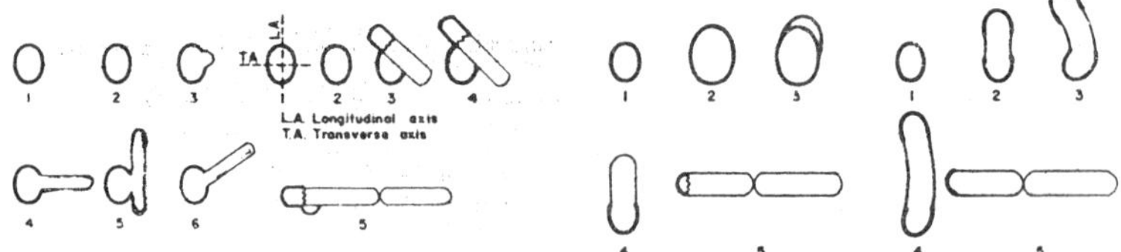

Fig. 6.5 Methods of spore germination. From left to right: equatorial germination without splitting along transverse axis, equatorial germination with splittng along transverse axis; polar germination; spore germination by comma-shaped expansion. (*After Lamanna.*)

II. Spore germination by absorption of the spore coat. Characteristics of this method are

A. The spore expands greatly during germination. A tripling or greater increase of the original volume occurs (**Fig. 6.6**).

B. The spore loses its characteristic refractiveness during germination, so that it is difficult to say when the spore has disappeared and the germ cell appeared.

C. After the second division of the germ cell, even if a thin capsule originally remains, all traces of the spore coat are gone.

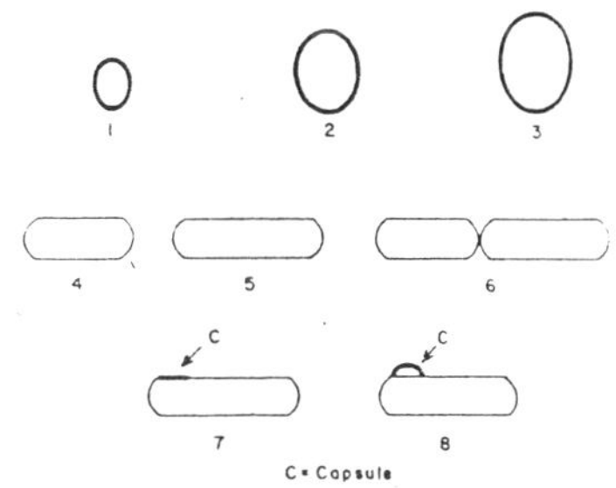

Fig. 6.6 Spore germination by absorption. (*After lamanna.*)

Some strains germinating by absorption regularly show a thin capsule remaining about one end of the growing cell. This would appear as a polar germination. In other cases, equatorial capsules are seen. Yet, in all instances, the spore is considered to germinate by absorption in as much as the three characteristics of the method are still adhered to.

7

Nutrition of Bacteria

Culture media (singular, medium) are solid, semisolid, and liquid nutrient preparations employed for the cultivation of micro-organisms. They are artificial environments prepared to simulate natural conditions as far as is necessary.

Strict Autotrophic Bacteria : The strict autotrophs cannot utilize organic matter and may even be harmed by its presence. These organisms are able to synthesize complex compounds composing their protoplasm from simple inorganic salts. They obtain their carbon form carbon dioxide and their energy from the oxidation of certain inorganic compounds or even elements. Because of this fact, they are independent of vegetable and animal life.

Strict Heterotrophic Bacteria : The strict heterotrophs cannot synthesize their complex protoplasm from simple inorganic salts but must have organic compounds, such as proteins, peptones, amino acids, and vitamins, for growth.

Facultative organisms : The facultative heterotrophic forms show characteristics intermediate between the two, being able to utilize both inorganic and organic compounds. They comprise the great majority of bacteria that have been studied and classified.

At one end of the scale, the organisms exhibit complete independence; at the other end, they show complete parasitism. Fildes (1934) advanced the theory that parasitism involved the loss of enzymes essential for the synthesis of bacterial protoplasm, making it necessary to add certain complex ingredients to the culture medium.

Common Ingredients of Culture Media

The ingredients commonly added to culture media and their uses are as follows:

Water : Water is absolutely necessary for the existence of living cells. Tap water many show considerable variation in composition from one locality to another, with the result that uniform culture media cannot always be prepared. The calcium and magnesium in tap water react with the phosphates present in peptones, beef extract, and other ingredients of culture media to give insoluble phosphates. The insoluble phosphates may not form in the cold, but during sterilization such media throw down considerable precipitate, which usually proves objectionable. Since distilled water is of definite composition. It should be used in preference to tap water for the preparation of culture media.

Peptones : Peptones are intermediate products of hydrolysis formed by the action of certain proteolytic enzymes (trypsin, pancreatin, papain, etc.) on native proteins. As hydrolysis proceeds, the large colloidal protein molecules become broken up into a series of smaller fragments which are designated, respectively, as proteoses, peptones, peptides, and finally amino acids. The proteoses till exhibit colloidal properties and it is customary to consider them as the last hydrolytic product still possessing true protein characteristics. In other words, the protein nature of the molecule disappears on further hydrolysis.

The commercial peptones are not the same as the peptone of the chemist, who uses the term in its narrow, chemical sense. The commercial preparations employed by the bacteriologist are composed of proteoses or albumoses, peptones, peptides, and amino acids. The proportions vary, depending upon the type of peptone. The usual commercial preparations contain a high percentage of peptones and amino acids, and only a negligible quantity of proteoses and other nitrogenous constituents. Others contain a higher content of proteoses with small amounts of peptones and amino acids. Still others contain all fractions in more or less well-balanced proportions. Some organisms prefer one type of peptone; others grow better on another type.

A whole protein, such as casein or egg albumin, is probably not attacked by bacteria when incorporated in a medium as the only source of nitrogen and carbon. The molecules are believed to be too large to enter a bacterial cell. If a trace of peptone is added to the medium, it will enter the cell and stimulate the elaboration of an extracellular proteolytic enzyme capable of attacking the whole protein. The protein fractions can then enter the bacterial cell, where they are acted upon by the intracellular enzymes.

The most important function of peptones in culture media is to furnish an available source of nitrogen. Since amino acids are amphoteric compounds, peptones are also excellent buffers.

Meat Extract : In its preparation, fresh, lean beef is cut into pieces, placed in a vessel with the appropriate amount of water, and heated for several hours with occasional stirring. The liquid portion is poured off and the solid material subjected to gentle pressure to separate any remaining liquid. The extract is cooled to remove fat, and strained. The clear liquid is evaporated in vacuo to the consistency of a pasty mass.

The constituents removed from muscle by boiling in water are known as extractives. The extractives obtained from fresh muscle tissue by boiling in water amount to about 2 per cent of the weight of the muscle. Two classes of extractives are obtained from meat: the nitrogenous and the non-nitrogenous. The nitrogenous extractives include creatine, xanthine, hypoxanthine, uric acid, adenylic acid, inosinic acid, carnosine, carnitine, glycocoll, urea, glutamine, β-alanine, etc. The non-nitrogenous extractives include glycogen, hexosephosphate, lactic acid, succinic acid, inositol, fat inorganic salts, etc.

The use of beef extract in culture media was introduced by Loeffler (1881) and has been a routine procedure in bacteriology ever since. Meat extract is added to media to supply certain substances that stimulate bacterial activity. It contains enzyme exciters, which cause accelerated growth of micro–organisms.

Mellwain et al. (1939) showed that glutamine, a constituent of beef extract, was a necessary nutrient for the growth of *Streptococcus pyogenes*. Williams (1941) believed that β-alanine was present in beef extract in small amounts, since it may arise from the hydrolysis of carnosine or in traces from pantothenic acid. Stokes, Gunness, and Foster (1941) analyzed beef extract for the presence of eight members of the B complex, namely, thiamine, riboflavin, pantothenic acid, nicotinic acid, biotin, pyridoxin, folic acid, and *p*-aminobenzoic acid, and found all of them to be present.

Yeast Extract : Yeast extract is prepared by extracting autolyzed yeast cells with water and evaporating the liquid to dryness in vacuo. The autolysis of the yeast cells is carefully controlled to prevent destruction of the natural-occurring vitamins of the B complex.

Yeast extract is an excellent stimulator of bacterial growth and is frequently used in culture media in place of meat extract. It is a rich source of the B vitamins and is used to supply these factors in culture media. For this reason it is superior to meat extract in most culture media.

Gelatin : Gelatin is a protein and is prepared by the hydrolysis of collagen with boiling water. Gelatin is not soluble in cold water but swells and softens when immersed in it. It is quite soluble in boiling water. On cooling, it solidifies to form a transparent gel.

Gelatin is rarely used ass a substitute for agar for the preparation of solid media because (1) it is attacked and decomposed by many bacteria and (2) it melts at 37°C. Gelatin is added to media principally to test the ability of organisms to liquefy it. Some organisms can liquefy it; others cannot. It is of importance in the identification of classification of bacteria.

Agar : Agar is the dried mucilaginous substance extracted from *Gelidium corneum* and other species of *Gelidium* and closely related algae. The plants are found growing chiefly along the coasts of Japan, China, Ceylon, Malaya, and Southern California.

Agar is the sulphuric acid ester of a linear galactan, insoluble in cold water but soluble in hot water, a 1 per cent neutral solution of which sets at 35 to 50°C. to a firm gel, melting at 80 to 100°C. Since agar is attacked by very few bacteria, it is used in preference to gelatin as a solidifying agent for the growth and isolation of bacterial species.

Stanier (1941, 1942) described seven well-recognized marine species belonging to the genera *Vibrio, Cytophaga,* and *Pseudomonas, and* several *Actinomyces* capable of liquefying agar. The enzyme responsible for agar digestion has been shown to be extracellular.

Sodium Chloride : Sodium chloride is commonly added to culture media to increase their osmotic pressures, although this is usually not necessary.

Sometimes media containing blood are required for (1) cultivation of bacteria or (2) the recognition of a hemolytic reaction on agar. Red blood cells are hemolyzed when added to water or to media having low osmotic pressures. Obviously such media cannot be used for the recognition of a hemolytic reaction. This may be prevented by the addition of sodium chloride in a concentration approximating that of an isotonic solution.

Sodium chloride does not act as a buffer. Salts such as phosphates and carbonates do possess a string buffering action and are frequently added to culture media for this purpose.

Inorganic Requirements : The inorganic requirements of bacteria are not well understood. The chief obstacle to work of this nature is the difficulty encountered in obtaining media sufficiently free of inorganic contaminants to permit accurate observations to be made. In the absence of accurate information, the following elements are usually supplied: Na, K, Mg, Fe, S, and P; while Cl, C, N, and H are usually obtained from organic matter.

Shankar and Bard (1952) reported the necessity of Ca, Mg, Fe, Na, and K for the growth of *Costridium perfringens* but not Zn, Mn, Co, or Cu. In the absence of Ca, the bacteria grew in an aggregated state, whereas Mg and K deficiency resulted in the appearance of filaments.

Webb (1948) aslo found Mg to be necessary for the same organism. Filamentous forms were produced in Mg-free media. Such filaments reverted to cells of normal morphology on subculture to a medium containing the metal. Similar results were reported by MacLeod (1951) in his work on *Streptococcus faecalis*, and Rochford and Mandle (1953) on *Diplococcus pneumoniae*. The latter workers believed the phenomenon was not dependent upon the presence of capsular material surrounding the cells but rather appeared related to incomplete separation of morphologic units.

MacLeod and Snell (1948) found that K was necessary for the growth of five lactic acid bacteria. The magnitude of the K requirement was greatly increased by the addition of Na and NH_4. whether or not these ions inhibited growth depended upon the ratio of their concentrations to that of K, and not upon the absolute amounts present.

Brown and Gibbons (1955) reported that Na, Mg, K, and Fe were essential for the growth of red halophilic (salt-loving) bacteria. In Mg deficient media the rod forms became coccoid in appearance. Growth did not occur in the complete absence of K in the medium.

Sulfur appears to be a universal constituent of living cells. Cowie, Bolton, and Sands (1950) found that the sulfate ion readily passed through cells of *E. Coli* and that the uptake of the element was directly proportional to cellular growth.

The iron requirements of bacteria have been studied by Waring and Werkman (1942, 1943, 1944). They employed an iron-deficient medium and found that cells of *A. indolognes* required a minimum of 0.025 part per million of iron in the culture medium for optimal growth. The organism grown on a medium without the addition of iron contained 0.0031 per cent iron. When grown in the same medium with the addition of its minimal requirement ((0.025 part per million), it contained 0.0073 per cent iron. When grown with a large excess, it contained 0.1049 per cent iron. The effect of iron deficiency on the enzyme systems of *A indologenes* showed that catalase, peroxidase, formic dehydrogenase, hydrogenase, and cytochrome systems were depressed.

Fermentable Compounds : Fermentable compounds serve two functions in culture media: (1) they furnish readily available sources of energy, provided the organisms elaborate the enzymes necessary to ferment the compounds; and (2) fermentation reactions are of great help in identifying and classifying organisms.

The fermentable compounds commonly added to culture media include:

Monosaccharides:
- Pentoses: arabinose, xylose, rhamnose.
- Hexoses: glucose, levulose, mannose, galactose.

Disaccharides:
- Sucrose, lactose, maltose, trehalose, melibiose,

Trisaccharids:
- Raffinose, melezitose, gentianose.

Polysaccharides:
- Starch, inulin, dextrin, glycogen, cellulose.

Alcohols:
- Trihydric: glycerol.
- Pentahydric: adonitol.
- Hexahydric: mannitol, dulcitol, sorbitol.

Glucosides:
- Salicin, amygdalin.

Non-carbohydrate compounds:
- Inositol.

Types of Culture Media

Culture media employed for the cultivation of micro-organisms may be divided into two groups on the basis of the nature of the ingredients entering into their composition: (1) synthetic media and (2) nonsynthetic media.

Synthetic Media : Synthetic media are composed of compounds of known chemical composition. They may be composed entirely of inorganic salts; or mixtures of inorganic salts and organic compounds, such as amino acids, lower fatty acids, hydroxy acids, alcohols, and carbohydrates; or inorganic and organic compound; with added vitamins. The exact chemical make-up of all ingredients is known so that two batches of the same medium can be duplicated to a high degree of accuracy. Synthetic media are employed where it is desired to ascertain what effect an organism will have on a certain compound. The nutritional requirements of bacteria may be accurately determined only by the use of synthetic culture media.

The literature shows an increasing number of synthetic media being recommended for the cultivation of all types of organisms, including the fastidious disease producers. Since the nutritional requirements of organisms are becoming better understood, it appears to be only a question of time until most culture media will be of the synthetic type.

Nonsynthetic Media : The non-synthetic media are composed of ingredients of unknown chemical composition. Some of these are beef extract, yeast extract, various peptones, meat infusion, blood, serum, and casein hydrolysate. It is practically impossible to prepare two identical lots of the same medium from different batches of the ingredients.

The usual culture media employed by the bacteriologist are of the non-synthetic type.

It is now known that the presence of growth factors in culture media is absolutely necessary of the successful cultivation of most bacteria. The failure of an organism to grow on a certain medium is probably due to the absence of one or more of the essential growth accessory substances. Media are usually selected for their ability to produce results rather than because they are known to contain the necessary growth substances. It seems highly probable that few media containing all the necessary accessory substances can be successfully employed for the cultivation of bacteria. Until such investigations are made, the bacteriologist will continue to employ many kinds of media, each more or less specific for a particular purpose.

Nutritional Requirements

Sources of Nitrogen : A number of recent studies have shown that many bacteria are able to utilize ammonium salts as their only source of nitrogen. This is true of the autotrophic and probably most of the heterotrophic bacteria. Other heterotrophs grow very poorly or not at all when ammonium salts are present as the sole source of nitrogen. These organisms require nitrogen in the form of amino acids. Since it is believed that the bulk of the amino acids in a medium are dissimilated with the liberation of ammonia, compounds furnishing nitrogen, whether amino acids or inorganic ammonium salts, are generally of equal value as immediately available sources of nitrogen.

The amino acid requirements of bacteria show wide differences. Essential amino acids for one organism may not be necessary for another. No rule can be formulated to apply to all bacteria. This must be determined for each organism.

Fildes, Gladstone, and Knight (1933) showed that the amino acid tryptophan was necessary for the growth of *Salmonella typhosa.* They found that the organism could derive its nitrogen requirement from a mixture of amino acids containing tryptophan, but ordinarily the organisms would not grow in the absence of this essential amino acid. Some strains of *S. typhosa,* which initially required tryptophan, could be trained to grow without it; others could not. They concluded that tryptophan was probably an essential constituent of protoplasm and that, if the organisms could not synthesize the compound, it must be added to the culture medium

Curcho (1948) succeeded in producing mutant cells of *S. typhosa* capable of growing in the absence of tryptophan. The mutants retained the tryptophan independence through daily transfer of over a year.

Rydon (1948), on the basis of experimental evidence, concluded that enthranilic acid was synthesized by *S. typhosa* and that the compound was a precursor of indole in the biosynthesis of tryptophan. He suggested the following scheme:

Anthranilic acid + Unknown substance →(A) Indole

Indole + Serine →(B) Tryptophan

Tryptophan + Other amino acids →(C) Protein

Fildes (1956) showed that strains of *S. Typhosa,* in which the indole → tryptophan enzyme was unimpaired, synthesized tryptophan in excess during growth. The amount produced might be so small as to be undetectable by the methods used, but when indole in excess was added to a culture, the production of tryptophan was much greater. Experiments with *E. Coli* also showed in accumulation of tryptophan in cultures containing indole. The survival of the tryptophan in this case was due to inhibition of the tryptophanase enzyme by the indole.

Use of Carbon Compounds : As indicated in the opening paragraphs of this chapter, the carbon requirements of organisms vary considerably. The autotrophic bacteria obtain their carbon from carbon dioxide or carbonates. The heterotrophic bacteria require carbon compounds for energy and cell synthesis. The variety of carbon compounds available to organisms varies considerably from one species to another, which probably accounts for the presence of organisms in specific environments.

Braun and Calm-Bronner tested a large number of carbon compounds, including formic, acetic, oxalic, lactic, succinic, malic, tartaric, and citric acids and glycerol, glucose, and arabinose. They found that glucose, glycerol, and lactic and citric acids were utilized more than any of the other carbon compounds when tested against *Salmonella schottmuelleri, S. enteritidis* and *S. Typhosa*. Acetic and oxalic acids ranked next. Formic and probably tartaric acids were not available as sources of carbon. The amino acids ranked lower than the organic acids, carbohydrates, and glycerol from the standpoint of availability. Den Dooren de Jong (1926) tested about 250 organic compounds for their availability as sources of carbon for a number of organism. The compounds were added to a synthetic medium containing ammonia as the only source of nitrogen. His conclusions were similar to those of Braun and Cahn-Bronner. He found that carbohydrates and related compounds were most generally utilized; these were followed by malic, citric, succinic, and lactic acids: next came the fatty acids; and last the monohydric alcohols.

Braun and Cahn-Bronner found that anaerobic growth was entirely absent when *S. schottmuelleri* was inoculated into an inorganic medium containing ammonium lactate and glucose. Koser (1923) found the same to be true when the members of the *Escherichia* and *Aerobacter* groups were inoculated into media containing various organic acids as carbon sources. Citric acid and its salts were utilized by *A. aerogenes* but not by *E. coli*.

Formation of Lipides : The formation of lipides by bacteria is dependent upon the nature of the carbon compounds added to media. Stephenson and Whetham (1922) employed an inorganic medium containing ammonium salts to which were added (1) lactic acid, (2) lactic and acetic acids, (3) glucose, and (4) glucose and acetic acid. The media were inoculated with *Mycobacterium phlei,* an acid fast organism. Their results are shown in Table 7.1. The addition of acetate to the various media produced no increase in protein formation, but did increase the lipide concentration. Increased concentrations of lactate and glucose increased both protein and lipide, usually the former. An organism that normally synthesizes sufficient lipide material to become acid-fast could be made to grow acid-sensitive by omitting from the culture medium an appropriate carbon source.

Table— 7.1 ***Effect of Composition of Medium on Fat Formation by Mycobacterium phlei***

Constituents of medium	*Period of maximum lipide formation, days*	*Milligrams per 100 ml. medium*		*Ratio, lipide/nitrogen*
		Nitrogen synthesized	*Lipides Synthesized*	
0.68% lactic acid	10	19	16	0.84
1.4% lactic acid	11	26	20	0.78
1.2% lactic acid 0.4% acetic acid	7	20	35	1.78
1% glucose	12	18	28	1.6
1% glucose 1% acetic acid	21	18	28	1.6
1% glucose	15	15	19	1.3
2% glucose	15	34	21	0.62

Larson and Larson (1922) reported that organisms which fermented glucose or glycerol did not synthesize additional lipide. On the other hand, organisms which did not ferment glucose or glycerol utilized the compounds for fat synthesis. Glycerol was superior to glucose for this purpose.

Geiger and Anderson (1939) inoculated *Agrobacterium tumefaciens* into two synthetic media, one containing glycerol and the other sucrose. The organisms grown on the glycerol-containing medium yielded only 2 per cent of total lipide, whereas those grown on the sucrose-containing medium gave 6 per cent of lipide. The nature of the fatty material synthesized by the organisms on the two media also showed considerable difference.

Table—7.2 ***Yield of Lipides from dried Agrobacterium tumefaciens***

	Medium 1, grams	*Medium 2, grams*
Bacteria used for extraction	364	276
Alcohol-ether-soluble lipides	4.85	16.23
Chloroform-soluble lipides	2.46	0.63
Total phosphatide	3.21	10.90
Total acetone-soluble fat	3.35	5.92
Ether-insoluble substance	0.64	0.03

Vitamins and Growth Factors

Hopkins (1906) was probably the first to point out that compounds other than fat, protein, carbohydrate, minerals, water, and oxygen are necessary in human nutrition. This observation led to the discovery of the vitamins by Funk (1912). Wildiers (1901) reasoned that, since certain factors not known at that time were required in human nutrition, a comparable situation existed in the requirements of fungi. He employed an inorganic medium containing cane sugar and ammonia as source of carbon and nitrogen, respectively, and found that yeast cells failed to grow unless a certain number of organisms (size of inoculum) were transferred to fresh medium. A larger inoculum carried more growth factors to the new medium, resulting in multiplication. The addition of a boiled suspension of yeast produced the same growth-promoting effect on small inocula as the addition of an emulsion of living yeast. Wildiers named the growth-promoting factor *bios*.

The term *growth factor* was originally applied to any compound that produced a stimulatory effect on the growth of organisms. The term *vitamin* was used in connection with any substance which in minute amount was necessary in the nutrition of animals. Since growth factors and vitamins have been shown to be the same, the terms now are used interchangeably. Williams (1941) coined the term *nutrilite* to include any organic substance, regardless of its nature, which in minute amount is of importance in the nutrition of an organism.

It is now known that bios is a mixture of several of the water-soluble members of the vitamin B complex, including thiamine (Vitamin B_1), pantothenic acid, pyridoxin (vitamin B_6), biotin, and inositol.

A large number of vitamins have been studied and their chemical make-up determined. Most of these have been synthesized. Some are simple in structure; others are quite complex.

Response of Bacteria to Growth Factors : The fact that an organism will grow in the absence of a particular growth factor in the medium does not mean necessarily that the factor is not required. Some organisms can synthesize one or more factors. Whereas others are unable to do so. Therefore, it is erroneous to conclude that an organism does not require a certain factor because it is absent from the medium in which the cells are growing.

O'Kane (1941) cultivated micrococci on a medium free of riboflavin and found that the organisms synthesized the growth factor. Burkholder and McVeigh (1942) demonstrated that certain intestinal species could synthesize a number of vitamins of the B complex. Actinomycetes were shown by Herrick and Alexopoulos (1943) to be capable of synthesizing thiamine. Miller (1944) reported that *Escherichia coli* synthesized folic acid. Altenbern and Ginoza (1954) showed that smooth cells of *Brucella abortus* possessed an enzyme which coupled pantoyl lactone and β-alanine to produce pantothenic acid.

Some bacteria can be adapted to dispense with certain vitamins by repeatedly subculturing to fresh media. For example, strains of *Propioni bacterium* can be adapted to grow without riboflavin and thiamine. Koser and Wright (1943) isolated variants of dysentery bacilli capable of growing without added nicotinamide.

Vitamins Required By Bacteria

The factors most frequently reported as promoting growth of bacteria include thiamine chloride, biotin, pantothenic acid, riboflavin, pyridoxin, nicotinic acid, and *p*-aminobenzoic acid. In addition to these, a number of miscellaneous compounds have been shown to be indispensable for some, but not necessarily for all, bacteria. Some of these are nicotinamide, inositol, pimelic acid, β-alanine, glutamic acid, glutamine, folic acid, purines, pyrimidines, glutathione, hematin, betaine, purine nucleotides, choline, and vitamin B_{12}.

Thiamine Chloride (Vitamin B_1) : This factor is probably necessary for the growth of all bacteria. A number of organisms have been shown to be capable of synthesizing the compound. Chemically it is a pyrimidine-thiazole compound having the following structural formula:

```
                               CH3
                               |
      N = C·NH2·HCl            C = C·CH2·CH2OH
      |   |                  /     |
 CH3·C    C·CH2—N                  |
     ||   ||    |  \\              |
      N—  CH    Cl    C  —  S
                      |
                      H
```

Thiamiine chloride

The pyrophosphate ester (thiamine pyrophosphate) is called cocarboxylase. It participates in all decarboxylations which lead to the formation of aldehydes and carbon dioxide. It has been shown that cocarboxylase could replace vitamin B_1 in the nutrition of yeasts. It is probable that vitamin B_1 serves as the precursor of cocarboxylase.

Biotin : This growth factor was first isolated by Kögl (1935). It is identical with vitamin H. It has been crystallized as pure biotin and as the biotin methyl ester.

Du Vigneaud et at. (1942) worked out the structure of the compound and showed it to be 2′ keto-3, 4-imidazolido-2-tetrahydrothiophene-n-valeric acid:

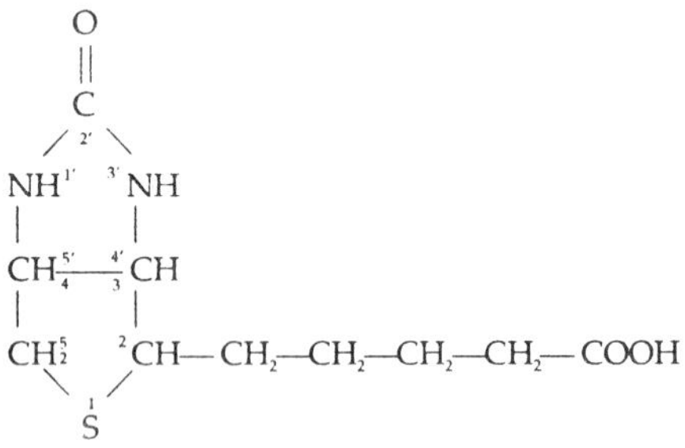

```
        O
        ||
        C
      / 2' \
  NH 1'    3' NH
  |            |
  CH 5'4 ——— 4'3 CH
  |            |
  CH 5 2      2 CH—CH2—CH2—CH2—CH2—COOH
      \      /
         S 1
```

Biotin

Biotin is present in all living cells in very minute amounts. The fungi (bacteria, yeasts, molds) are the best sources of the vitamin. According to Kögl, an amount as small as 0.00004 microgram added to 2 ml. of a yeast culture caused a 100 per cent increase in growth. This is equivalent to 1 part in 50 billion parts of medium.

The exact function of biotin is not clearly understood. It appears to participate in various enzymatic reactions. Shive and Rogers (1947) believed that biotin functioned as a coenzyme in the carboxylation of pyruvic acid to oxalacetic acid:

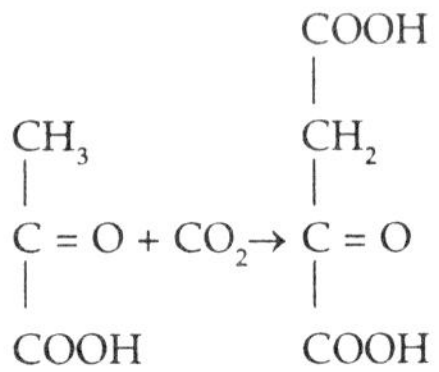

$$\begin{array}{c} CH_3 \\ | \\ C=O \\ | \\ COOH \end{array} + CO_2 \rightarrow \begin{array}{c} COOH \\ | \\ CH_2 \\ | \\ C=O \\ | \\ COOH \end{array}$$

Pyruvic acid Oxalacetic acid

The oxalacetic acid, in the presence of glutamic acid or alanine, may be converted into aspartic acid.

Stokes, Larson, and Gunness (1947) found that biotin could completely substitute for aspartic acid in the growth of a number of bacteria. They concluded that biotin participated in the synthesis of aspartic acid.

Delwiche (1950) stated that biotin was closely concerned with the decarboxylation of succinic acid to propionic acid by *Propionibacterium pentosaceum.*

Lichstein (1950) isolated a coenzyme from yeast extract which contained an acid-stable material that replaced biotin for growth of *Saccharomyces cerevisiae.* He believed that this acid-stable component was a derivative of biotin that might be an intermediate in the synthesis of the coenzyme from biotin. This coenzyme was active in the oxalacetate decarboxylase and succinic acid decarboxylase systems.

Carlson and Whiteside-Carlson (1949) found that *Leuconostoc* did not require biotin for the utilization of sucrose, but that the vitamin was essential for growth when the constituent monosaccharides were added to the medium.

Campbell and Williams (1953) reported that all biotin-requiring strains of *bacillus coagulans* and *B. stearothermophilus* could substitute oxybiotin. Some strains could satisfy their biotin requirement with aspartic acid; some with oleic acid; and some with pimelic acid.

Pantothenic Acid : Williams and Bradway (1931) were the first to show that a growth factor which they called pantothenic acid was necessary in the nutrition of yeast. In a later report, Williams et al. (1933) stated that pantothenic acid was a growth determinant of universal biological importance, being present in all living cells. It was shown to consist of β-alanine united to a saturated dihydroxy acid by a peptide-like combination:

$$
\begin{array}{cccccccccccccc}
 & & CH_3 & & OH & & O & & H & & H & & H & \\
 & & | & & | & & \| & & | & & | & & | & \\
HOH_2C & - & C & - & C & - & C & - & N & - & C & - & C & - COOH \\
 & & | & & | & & & & & & | & & | & \\
 & & CH_3 & & H & & & & & & H & & H &
\end{array}
$$

Pantothenic acid

Pantothenic acid is a component of coenzyme A, which is involved in the acetylation of aromatic amines and choline. Pantothenic acid is also related to the utilization of other vitamins, especially riboflavin.

Riboflavin : This factor has been studied extensively and found to be necessary for the growth of a large number of organism. It is sometimes referred to as vitamins B_2, G, lactoflavin (from milk), and ovoflavin (from eggs). chemically it is 6, 7-dimethyl-9-(D-1′-ribityl) isoalloxazine:

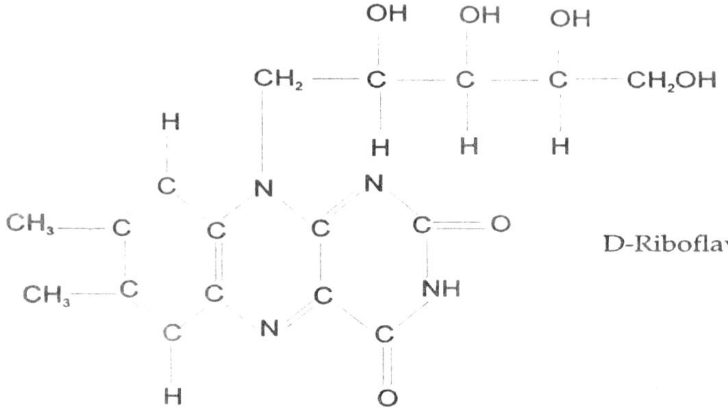

OH OH OH
CH_2 — C — C — C — CH_2OH
H H H H
C N N
CH_3 — C C C C = O
CH_3 — C C C NH
C N C
H O

D-Riboflavin

Riboflavin pays an important function in many enzyme systems. It is a component of a number of enzymes known as the flavoproteins. These include Warb'urgs yellow enzyme, diaphorase, cytochrome cytochrome c reductase, xanthine dehydrogenase, D-amino acid oxidase, etc.

Pyridoxin : This growth factor is sometimes called vitamin B_6. It was first recognized as a vitamin by György (1935), Keresztesy and Stevens (1938) and Lepkovsky (1938), working independently, isolated the vitamin in crystalline form. Harris and Folkers (1939) synthesized the vitamin and showed it to be

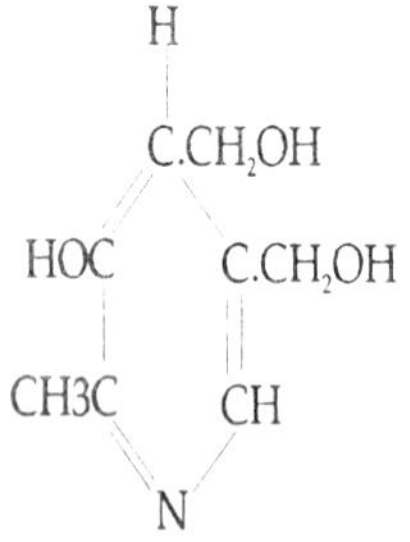

H
$C.CH_2OH$
HOC $C.CH_2OH$
CH3C CH
N
Pyridoxin

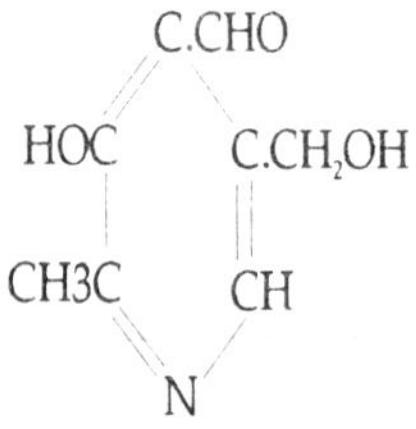

$C.CHO$
HOC $C.CH_2OH$
CH3C CH
N
Pyridoxal

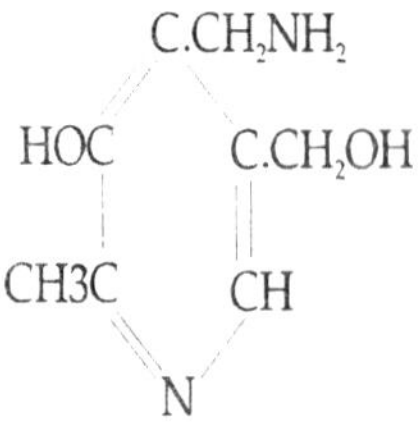

$C.CH_2NH_2$
HOC $C.CH_2OH$
CH3C CH
N
Pyridoxamine

2-methyl-3-hydroxy-4, 5-(di-hydroxy methyl) - pyridine. Two other important naturally occurring substances with vitamin B_6 activity are pyridoxal, 2-methyl-3-hydroxy-4formyl-5-hydroxy-methylpyridine, and pyridoxamine, 2-methyl-3-hydroxy-4-aminomethyl-5-hydroxymethylpyridine:

These three vitamins function in the form of a coenzyme. This co-enzyme is pyridoxal-5-phosphate; it is also called codecarboxylase, cotransaminase, etc. Pyridoxin, pyridoxal, and pyridoxamine owe their vitamin activity to the ability of the organism to convert them into the active pyridoxal-5-phosphate. It functions as a transaminase for the synthesis of amino acids from their keto analogues. Pyridoxal-5-phosphate has also been reported to function as a glutamate-asparatate transaminase. When combined with protein (apoenzyme), pyridoxal-5-phosphate catalyzes the decarboxylation of amino acids.

Nicotinic Acid (Niacin) and Nicotinamide (Nicotinamide) (Niacinamide). Nicotinic acid was first prepared in 1867, but the nutritional importance of the compound was not recognized until 1937 by Elvehjem et al. The structural formulas of nicotinic acid and its amide are as follows:

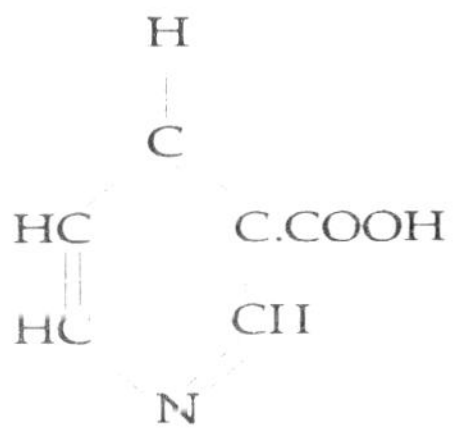

H
C
HC C.COOH
HC CH
N

Nicotinic acid

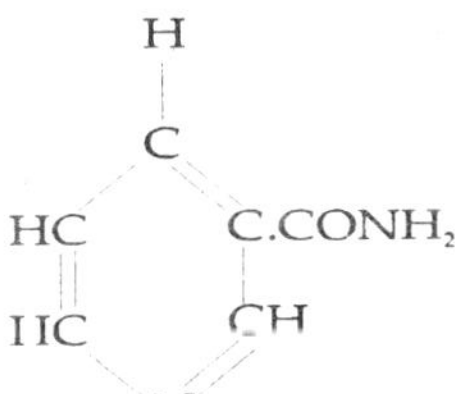

H
C
HC $C.CONH_2$
HC CH
N

Nicotinamide

Nicotinic acid or its amide is required by all living cells. It is an essential part of coenzymes I and II. As far as it is known, nicotinic acid and it derivatives have no function in living cells other than as parts of the above-named coenzymes.

p-Aminobenzoic Acid. Woods (1940) and Fildes (1941) showed that *p*-aminobenzoic acid (PABA) was highly active in reversing the bactèriostatic action of the sulfonamides. They reported that sulfanilamide competed with PABA for an enzyme and thus interferred with some essential metaboic reaction.

Landy et al. (1943) found that sulfonamide-resistant strain of *staphylococcus aureus* produced greater amounts of PABA than their parent strains. The quantity of PABA synthesized by resistant strains appeared sufficient to account for their resistance to sulfonamide drugs. It has the following structural formula:

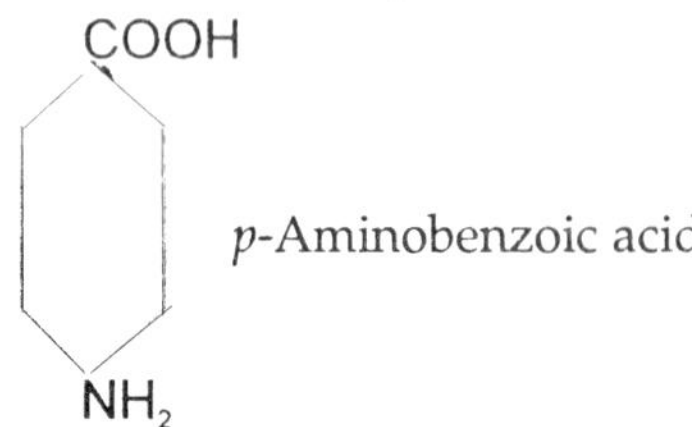

PABA is a unit in the structure of another vitamin, namely; pteroyglutamic acid or folic acid. It is almost universally agreed that the sulfonamides act by inhibiting the conversion of PABA to folic acid.

Folic Acid : Snell and Peterson (1939, 1940) found that *Lactobacillus casei* required a hitherto unrecognized growth substance, which they called the *L. casei* eluate factor. They noted that the factor possessed many of the properties of a purine. Stokstad (1941) found that a mixture of guanine and thymine possessed some of the growth-promoting properties of the eluate factor. Mitchell, Snell, and Williams (1941) concentrated the factor to a state approaching purity and named it *folic acid.*

Chemically, folic acid is pteroylglutamic acid, having the following structure:

N N

COOH HC C $C{\cdot}NH_2$

$HOOC{\cdot}CH_2{\cdot}CH_2{\cdot}CHNH{\cdot}CO$ — (benzene ring) — $NH{\cdot}CH_2{\cdot}C$ C N

N C

OH

Glutamic acid | p-Aminobenzoic acid | 2-Amino-4-hydroxy-6 methylpteridine

Pteroic acid

p-Aminobenzoic acid and glutamic acid are components of folic acid. Folic acid appears to be primarily concerned with the synthesis of purines and pyrimidines. It is also believed to be a requirement for the synthesis of certain amino acids which involved the incorporation of a single carbon fragment. It appears that carbon dioxide may serve as the source of the carbon fragment in this experimental system.

Inositol : This vitamin was the first pure substance isolated that was found to contribute to bios activity. Inositol was first isolated in pure form by Eastcott (1928) from yeast.

Nine stereoisomeric forms of inositol are possible. Of these, seven are optically inactive or *meso* forms, and two are asymmetric enantiomorphs. The form which is most widely distributed in nature and which is of importance in the nutrition of micro–organism is optically inactive. It has been named *myo*-inositol, and has the following structure:

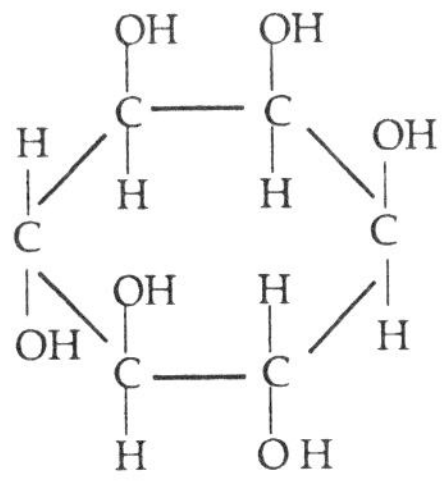

Inositol

Inositol is of widespread occurrence and great abundance in nature. Because of this fact it is believed to act both as a vitamin and as an energy-yielding nutrient. Its exact function in living cells in terms of enzymes and coenzymes is not clearly understood. It has been shown to be a constituent of certain phospholipides isolated from bacteria. It combines with phosphate, proteins, fatty acids, glycerol, and galactose.

Vitamin B_{12} : This is a cobalt and phosphorus-containing vitamin, usually produced by microbial fermentation or obtained from liver.

Evidence seems to indicate that the vitamin is a polypyrrole related in some way to hemin or to bile pigments. Its exact chemical structure is not clearly understood.

Vitamin B_{12} affects the phosphorus metabolism of *Lactobacillus leichmannii*. The vitamin increases the uptake of phosphorus in the desoxyribonucleic acid' fraction of the cell, which would indicate that it is involved in nucleic acid synthesis.

Chaplin and Lochhead (1956) studied the B_{12} requirements of a species of the genus *Arthrobacter*. They showed that with suboptimal concentrations of the vitamin, growth in liquid medium was flocculent, whereas cultures with adequate amounts were uniformly turbid. Flocculation was associated with abnormal cell morphology. The cells were noticeably swollen, elongated, and irregularly bent and showed rudimentary branching.

Pimelic Acid and β-Alanine : Mueller (1935) and Mueller and Kapnick (1935) showed that the diphtheria bacillus, *Corynebacterium diphtheriae,* produced a luxuriant growth in a medium composed entirely of amino acids to which were added a purified liver extract and a carbon source, such as glycerol and lactic acid, and suitable inorganic salts. Mueller (1937a, b) identified two of the constituents of the liver extract, essential for growth of the diphtheria bacillus, as pimelic acid and nicotinic acid. In a later report, Mueller and Cohen (1937) found *β*-alanine to be a third growth accessory substance present in liver extract responsible for the luxuriant growth of the diphtheria bacillus in a synthetic medium. The structural formulas for pimelic acid and *β*-alanine are as follows:

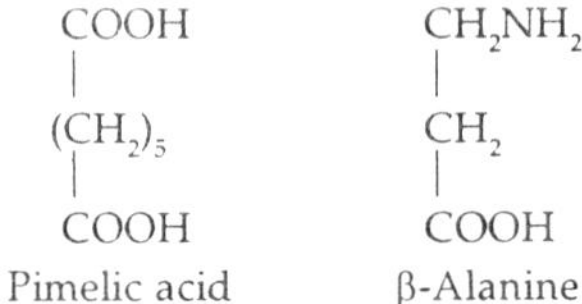

Pimelic acid is one of the degradation products of biotin and has been shown to be capable of serving as a biotin precursor for some micro-oganisms.

β-Alanine is a component of pantothenic acid, being united to a saturated dihydroxy acid by a peptide-like linkage.

Glutamic Acid and Glutamine : McIlwain et al. (1939) found that glutamine was an essential nutrient for the growth of some strains of *Streptococcus pyogenes* but not for others. Glutamine is present in beef extract and is probably widely distribuited in the animal body. Later Fildes and Gladstone (1939) and McIlwain (1939) reported that glutamine was indispenable for the growth of many strains of *S. pyogenes* and other bacterial species. Pollack and Lindner (1942) stated that nine species of lactic acid-producing bacteria required either glutamine or glutamic acid for growth. They concluded that bacteria required glutamine or glutamic acid for the construction of cell proteins because the requirements of these amino acids are of the order of magnitude that would be expected for this function. Lankford and Snell (1943) found glutamine to be necessary for the growth of certain strains of *Neisseria gonorrhoeae*. Glutamic acid is a component of another vitamin, namely, folic acid.

The structural formulas for glutamic acid and glutamine are as follows:

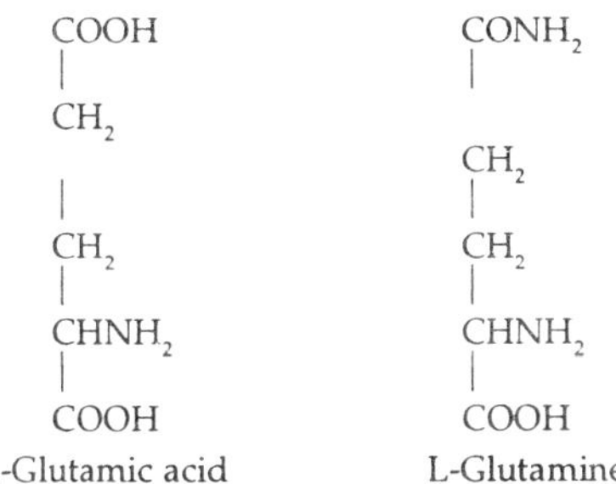

Purines and Pyrimidines : Richardson (1936) found that uracil was necessary for the growth of *Staphylococcus aureus*. Snell and Mitchell (1941) reported the necessity of uracil for the growth of *Lactobacillus Plantarum*, thymine for *Streptococcus lactis*, and adenine for *L. plantarum*. Shull, Hutchings and Peterson (1942) and Pollack and Lindner (1942) found that adenine, guanine, uracil, and xanthine were necessary for the growth of *L. casei*, Hutner 1944) and Rogers (1944) reported the necessity of uracil for *Shigella paradysenteriae* and *S. Pyogenes*, respectively.

Chattaway (1944) showed that orotic acid stimulated the growth of *L. casei*. Wieland et.al. (1950) reported the necessity of orotic acid for *L. bulgaricus*. Wright et al. (1950, 1953) showed that ureidosuccinic acid could partially replace the orotic acid requirement, and that both were precursors of nucleic acid pyrimidines. Snell, Kitay, and McNutt (1948); Wright, Skeggs, and Huff (1948); and Kitay, McNutt, and Snell (1950) found that desoxyribosides stimulated growth of lactic acid bacteria. Pearson (1949) reported the need of a number of purines for the growth of *Photobacterium fischeri*.

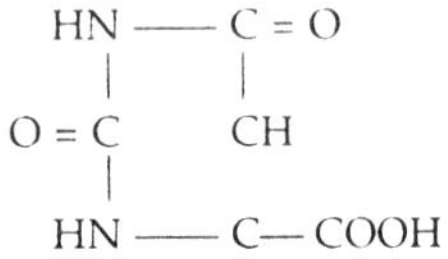

```
HN —— C = O
 |     |
O = C   CH          Orotic acid
 |                  (Uracil-4-carboxylic acid)
HN —— C— COOH
```

It is quite likely that purines are utilized in the synthesis of nucleic acids and related substances. Since folic acid contains a purine-like component, small amounts are probably utilized in the synthesis of that vitamin.

V and X Factors : Pfeiffer (1893) reported that the organism *Haemophilus infuenzae* would not grow in a broth medium unless blood was added. Thjötta (1921), Thjötta and Avery (1921*a*, *b*), and Fildes (1921, 1922) found that the same organism required the presence of two factors which they named the *V* and *X* factors, both of which are present in blood. Thee *V* factor is also present in many plant extracts and in a large number of bacteria. It is thermolabile, being destroyed in 15 min. at 90°C., is very sensitive to alkali but not to acid, diffuses through parchment membranes, and is not readily destroyed by atmospheric oxygen. The X factor is found in potatoes and in some bacteria. It is thermostable, resisting a temperature of 120°C. for 45 min.

H. influenzae is unable to grow on media containing only the X factor. However, it will grow on media in association with an organism, such as Staphylococcus *aureus,* which is capable of producing the V factor. The characteristic arrangement of colonies of *H. influenzae* in such an association is sometimes referred to as the satellite phenomenon. The hemophilic organism grow as satellites in isolated colonies at some distance from the colonies of *S. aureus*.

Lankford et al. (1943) observed the same phenomenon on agar plates streaked with a mixture of nonpigmented *Micrococcus* and *Neisseria gonorrhoeae*.

The organism *H. parainfluenzae* requires only the V factor for growth. Lwoff and Lwoff (1937*a*, *b*) isolated the V factor from yeast and added the extract to a culture of the organism. They noted that V activity paralleled the codehydrogenase I concentration of the yeast extract. This substance was found to replace the V factor in extremely low concentration. Codehydrogenases I and II are very similar chemically but are not, as a rule, interchangeable. As growth factors, however, codehydrogenase II can replace codehydrogenase I. Since one or the other factor must be supplied before growth can occur, they are considered to be true vitamins. Chemically, codehydrogenase I is diphosphopyridine nucleotide and codehydrogenase II is triphosphopyridine nucleotide.

X factor requirements are supplied largely by the addition of hemin. Since the addition of a small amount of blood enhances activity of the organism, it is quite likely that other factors are also involved.

Putrescine. This putrefactive compound has been shown to be an essential growth factor for *Haemophilus parainflunzae* and *Neisseria perflava*.

Chemically, putrescine is tetramethylene diamine and may be produced by the decarboxylation of ornithine:

$$\underset{\text{Ornithine}}{NH_2(CH_2)_3.CHNH_2.COOH} \rightarrow \underset{\text{Purescine}}{NH_2(CH_2)_3.CH_2NH_2} + CO_2$$

Ornithine in turn may be produced by the hydrolysis of arginine.

Glutathione : Gould (1944) reported that glutathione was necessary for the growth of *Neisseria gonorrhoeae*. The factor was found to be present in meat infusion, yeast infusion, and red-blood-cell extract. Freshly isolated strains of the organism were found not to require gutathione, but they showed a tendency to develop dependence on glutathione after some weeks of subculturing on a medium containing meat infusion.

Glutathione possesses several functions: (1) it acts as a carrier of hydrogen; (2) it prevents inactivation of sulfhydryl groups of enzyme; and (3) it is believed to maintain ascorbic acid in the reduced form.

Effect of pH on Growth-factor Requirements : Doede (1945) found that the pH of the medium had a marked influence upon the growth-factor requirements of a number of bacterial species. The amount of nicotinic acid required by *Staphylococcus aureus* for maximum growth was approximately fifteen times greater at pH 8.0 than at pH 6.0 *Shegella paradysenteriae* required nicotinic acid in the medium at pH 7.0 and 8.0 but not at pH 6.0. At pH 6.0 the organisms were capable of synthesizing the factor. *Lactobacillus casei* grew in a medium without pyridoxin at pH 5.0 but failed to grow at pH 6.0 or 7.0 unless supplied with the factor. Folic acid, riboflavin, biotin, nicotinic acid, and pantothenic acid were found to be most effective at pH 6.0.

Bacteria Destruction of Vitamins : A number of bacterial species have been shown to be capable of destroying growth factors. Young and Rettger (1943) found that vitamin C (ascorbic acid) was easily destroyed by the enteric bacteria, including the intestinal streptococci. In the presence of an easily fermentable carbohydrate, such as glucose, the vitamin was protected from decomposition, whereas in the absence of the competitive agents, the ascorbic acid content of the medium became rapidly depleted.

Foster (1944 *a*, *b*) showed that the organism *Pseudomonas riboflavinus* was capable of oxidizing riboflavin to lumichrome according to the reaction.

$$\underset{\text{Riboflavin}}{C_{17}H_{20}O_6N_4} + 5\tfrac{1}{2}O_2 \rightarrow \underset{\text{Lumichrome}}{C_{12}H_{10}O_2N_4} + 5CO_2 + 5H_2O$$

Koser and Baird (1944) reported the destruction of nicotinic acid by bacteria of the *Pseudomonas fluorescens* and *Serratia marcescens* groups. Destruction occurred during periods of active cell multiplication. Metzger (1947) showed that members of the genus *Pseudomonas* utilized pantothenic acid as a growth factor. However, during growth these organisms destroyed 100 per cent of the vitamin within 72 hr. by a process of oxidation.

Bacteria Synthesis of Vitamins : Mayer and Rodbart (1946) reported the synthesis of riboflavin by *Mycobacterium smegmatis* on a synthetic medium. Smith and Emmart (1949) found the same to be true for the tubercle bacillus. Tanner, Vojnovich, and Van Lanen (1949) employed an organic medium and reported the formation of riboflavin by *Ashbya gossypii.*

Vitamin Content of Ingredients of Culture Media : Information concerning the vitamin content of ingredients of culture media should prove of interest and value in the cultivation of micro–organism. Such information could be used to decide whether a particular culture medium satisfies the growth-factor requirements of an organism.

Stokes, Gunness, and Foster (1944) assayed a number of culture media for their content of eight members of the B complex: thiamin, riboflavin, pantothenic acid, nicotinic acid, biotin, pyridoxin, folic acid, and *p*-amino-benzoic acid, and compared the results with the vitamin requirements of various micro–organisms that were unable to synthesize these factors. They arranged the various culture-media ingredients in order of descending value on the basis of their content of vitamins of the B complex.

The classification is as follows:

1. Yeast extract.
2. Meat extract, brain infusion, heart infusion.
3. Various peptones.

With the possible exception of thiamine, yeast extract is an excellent source of all vitamins of the B complex. This explains why yeast extract is held in such high favour as an ingredient of culture media.

They concluded that if peptone, meat extract, etc., were used in concentrations of 1 or 2 per cent, singly or in combinations, the resultant media were likely to be deficient in thiamine, riboflavin, pantothenic acid, pyridoxin, and *p*-aminobenzoic acid, but not in nicotinic acid, biotin, and folic acid. They suggested that such growth-factor deficiencies in culture media could be remedied by proper combination of ingredients or by the addition of yeast concentrates or synthetic vitamins.

Metabolite Antagonists (Antimetabolites) : A metabolite may be defined as any substance essential to growth and reproduction of cells. It may be synthesized by the cells or obtained from the environment,. A metabolite antagonist is any substance that blocks or inhibits the normal function of the metabolite. The best examples of antimetabolites involve substances structurally related to the substrate.

Inhibition of Succinic Dehydrogenase. The enzyme succinic dehydrogenase is inhibited by malonic acid:

$COOH{\cdot}CH_2{\cdot}COOH$	$COOH{\cdot}CH_2{\cdot}CH_2COOH$
Malonic acid	Succinic acid

The homologous dibasic malonic acid is believed to inhibit the enzyme by competing with the normal metabolite succinic acid. The competitive inhibition between the two compounds results almost in a complete blocking of the enzymic reaction.

Inhibition of p-Aminobenzoic Acid. the close structural reationship between *p*-aminobenzoic acid and sulfanilamide may be shown by the following formulas:

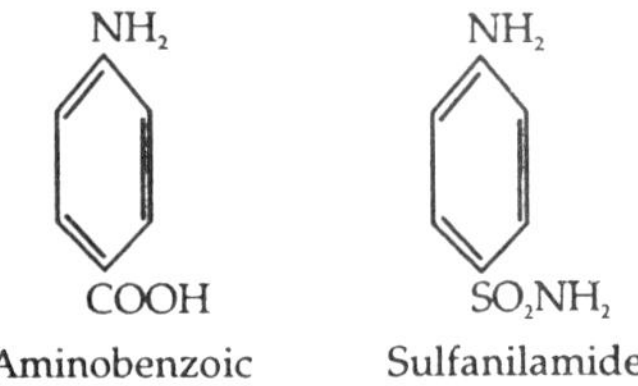

p-Aminobenzoic Sulfanilamide

Woods (1940) showed that *p*-aminobenzoic acid (PABA) in high dilutions antagonized the action of sulfanilamide. Fildes reported that PABA is an essential metabolite normally associated with any enzyme. Sulfanilamide displaces PABA from its enzyme, thereby stopping this essential line of metabolism. Other sulfa drugs (sulfonamides) behave in a similar manner.

Inhibition of Arginine : Volcani and Snell (1948) showed that arginine was an essential metabolite for the growth of several species of lactic acid bacteria. None of the species tested could use canavanine in place of arginine:

```
  NH2                                        NH2
 /                                          /
NH=C                                  NH=C
 \                                          \
  N·CH2 (CH2)2·CHNH2·COOH                  N—O—(CH2)2·CHNH2·COOH
  H                                          H
```

For some of the bacteria tested, canavanine functioned as an effective growth inhibitor. This was true both for organisms which synthesized arginine and for those which required this amino acid preformed.

Growth Phases in a Culture

Most bacteria multiply at a very rapid rate and produce pronounced changes in culture media in a short period of time. Under favourable conditions a single cell of *Escherichia coli* divides into two about every 20 min. If this same rate is maintained, a single organism will give 1 billion new cells after a period of about 10 hr. However, this rate of multiplication is not maintained indefinitely, owing to the exhaustion of the nutrients, to the accumulation of toxic metabolic waste products, and to the fact that many of the cells die. The rate of death increases as the culture ages. The more vulnerable cells die first, leaving the resistant forms in the culture at the end of the incubation period.

When an organism is inoculated into a tube of medium such as nutrient broth, multiplication does not take place in a regular manner. On the contrary, various growth phases may be recognized which are known as the life phases of a culture. Buchanan (1918) recognized seven distinct cultural phases which he designated as follows:

1. *Initial Stationary Phase.* During this phase, the number of bacteria remains constant. Plotting the results on graph paper gives a straight horizontal line (1*a*) in Fig. 7.1.
2. *Lag Phase or Phase of Positive : Growth Acceleration :* During this phase, the rate of multiplication increases with time (*ab*).
3. *Logarithmic Growth Phase :* During this phase, the rate of multiplication remains constant (*bc*). This means that the generation time is the same throughout.
4. *Phase of Negative Growth Acceleration :* During this phase, the rate of multiplication decreases (*cd*). The average generation time increases. The organisms continue to increase in numbers but at a slower rate than during the logarithmic growth phase.

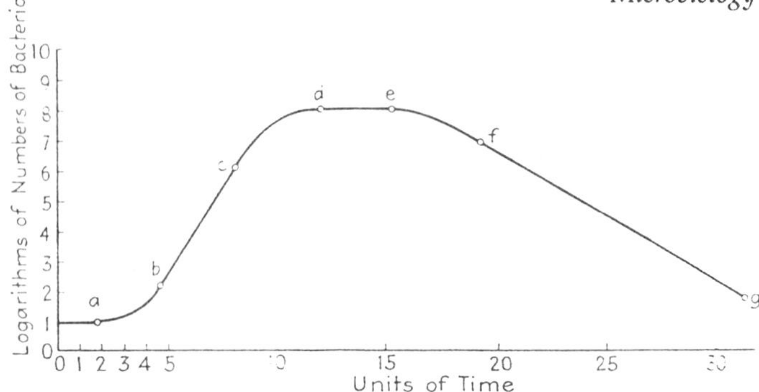

Fig. 7.1 Growth phases in a culture. (After Buchanan.)

5. *Maximum Stationary phase* : During this phase, the number of living organisms remains constant, i.e., the death rate equals the rate of increase (*de*)
6. *Phase of Accelerated Death* : During this phase, the numbers fall off with increasing rapidity (*ef*). The average rate of death per organism increases to a maximum.
7. *Logarithmic Death Phase* : During this phase, the rate of death is constant (*fg*).

It is generally believed that the phases in a bacterial culture are due to changes in the environment, such as alteration in pH, exhaustion of food supply, and accumulation of waste products.

Lag Period : When bacteria are transferred from an old culture to a new medium, they exhibit a period of delayed multiplication or lag. Müller (1895) was probably the first to make this observation. He recognized three distinct phases in a culture, which he designated the lag, logarithmic increase, and the slackened growth phases. Müller found that when cultures of differing ages were used for the inoculation of new medium the generation times in the new cultures showed considerable differences. Transfers from a typhoid culture 2½ to 3 hr. old gave a generation time of 40 min. in the new medium; a culture 6 $\frac{1}{4}$ hr. old gave a generation time of 80 to 85 min.; and a culture 14 to 16 hr. old gave a generation time of 160 min.

Another characteristic of organisms is that they show considerable variation to harmful influences in the different growth phases. Cultures in the lag and early logarithamic phases exhibit greater sensitivity to heat than these in the older phases of growth. Schultz and Ritz (1910) reported that a 20-min. culture of *E. coli* was more resistant to heat than a 4-hr. culture. Then the resistance showed a steady rise as the culture aged. Sherman and Albus (1923) exposed *E. coli* to various unfavourable conditions and found old cells to be considerably more resistant than very young cells.

If a tube of fresh medium in inoculated from a culture in the logarithmic growth phase, **(Fig. 7.1)** the lag phase will be greatly reduced and in many cases completely eliminated. The organisms in the logarithmic growth phase are multiplying at the maximum rate and continue to do so when transferred to fresh medium. Buchanan (1928) concluded that transfer from any phase of a bacteria culture cycle to a new medium is followed by a continuance of the phase of the parent culture.

Factors Affecting Cell Size : Henrici (1928) showed that organisms increased very markedly in size during the lag phase. He found that the average length of *Bacillus megaterium* was six times that of the inoculated organisms taken from an old culture. This increase generally manifested itself after 2 hr., and the maximum size was usually noted between 4 and 6 hr. During the lag phase, the cells showed considerable fluctuation in form. On passing from the lag phase to the logarithmic death phase, the organisms gradually decreased in size and exhibited a more constant cell form.

It is well known that when a parent culture is inoculated into a new medium, an initially slow rate of increase in bacterial numbers occurs. This slow rate of multiplication cannot be interpreted as indicating a period of lag in the sense of decreased viability and activity. What actually happens is that the rate of multiplication decreases but the individual cells become larger, giving a rapid increase in bacterial mass. During the first 2 or 3 hr., new culture inoculated from 3-hr. old parent cultures show a slower multiplication rate than new cultures inoculated from 24-hr. parent culture, but the increase in protoplasmic growth remains the same. The cells from a young parent culture show the same increase in cell mass as the cells from an old parent culture even though they multiply at a slower rate. The rate of increase in cell mass is nearly constant from the time growth first begins until the maximum population is reached.

The results suggest that conditions in fresh medium favour an increase in cell size but inhibit cell division, with the result that a majority of the cells attain an abnormal size before fission occurs. Inoculation of a large number of cells into new medium tends to produce the reverse effect, i.e., the average size is smaller and attained much sooner than with small inoculations. If cells are removed from a culture before their average maximum size is reached and are transferred to fresh medium, the organisms attain a larger size and the critical point takes place later than in the case of the original culture. This would indicate that the size of the organisms is dependent upon the density of the bacterial culture. The concentration of the nutrients in the medium is another factor. The maximum size of the organisms in a dilute medium is smaller and the critical point is reached earlier. A dilute medium produces a poorer growth than a more concentrated preparation. This mean that a more concentrated medium showing a heavy, crowded growth produces the same effect on cell size as a more dilute medium showing a light growth.

Effect of Carbon Dioxide on Growth : In general, bacteria grow better in the presence of an increased concentration of carbon dioxide. Walker (1932) found that the length of the lag phase could be controlled by the concentration of carbon dioxide present. He noted that the multiplication of *E. coli* could be delayed indefinitely by aeration of the culture with carbon dioxide-free air. Reintroduction of carbon dioxide into the medium caused a rapid increase in bacterial numbers. He concluded that "the phenomenon of lag may be due largely, if not entirely, to the time it takes the culture to build up the CO_2 content of the medium or of the cells themselves to a value essential for growth." Others have come to a similar conclusion. There appears to be no doubt that the amount of carbon dioxide present in a new medium is an important factor in controlling the length of the lag phase, but it is probably not the only factor involved.

Factors Affecting Rate of Reproduction : The rate of multiplication of bacteria is increased by a rise in temperature.

This continues until a certain maximum is reached, after which the rate decreases until death finally occurs. The generation times of *E. coli* at different temperatures of incubation are as follows:

°C.	*Time*	°C.	*Time*
10	14 hr., 20 min.	35	22 min.
15	120 min	40	17½ min.
20	90 min.	45	20 min.
25	40 min.	47½	77 min.
30	29 min.		

The composition of the medium also affects the generation times of bacteria. The generation time of *S. typhosa* in a 1 per cent solution of peptone is about 40 min. at 37°C. If the amount of peptone is less than 0.2 per cent, the generation time is almost inversely proportional to its concentration; if more than 0.4 per cent, an increase in the peptone concentration is practically without effect on the growth rate of the organism. The addition of 0.175 per cent of glucose to a medium containing only 0.1 per cent peptone lowers the generation time from about 111 to 50 min. The addition of the same amount of glucose to a 1 per cent peptone solution reduces the generation time only from about 39 to 34 min.

Effect of Age on Cell Morphology : Under some environmental conditions, bacteria show the presence of granules; under other conditions, they do not. It has been observed that when cells are largest (2 to 4 hr. old), intracellular granules disappear and the protoplasm becomes more hyaline and stains more deeply. As the cells age and decrease in size, they become increasingly more granular. Old cells are, in general, very granular, whereas young cells do not exhibit the presence of granules. An exception to this rule is the organism *Corynebacterium diphtheriae,* the causative agent of diphtheria. This organism appears to be smaller in young than in old cultures and to exhibit the presence of granules in both young and old cultures and to exhibit the presence of granules in both young and old cells.

Effect of Constant Environment on Cell Numbers : Jordan and Jacobs (1944) cultivated *E. coli* in an apparatus that permitted rigid control of temperature, pH, aeration, and culture volume and allowed food to be supplied at a constant rate by means of an automatic syringe mechanism. Determinations of total and viable cell populations were made. Results showed an initial period in which the total and viable counts both increased, followed by a steady phase in which the viable count remained constant or decreased slightly, while the total cell counts steadily increased. When during the steady phase the food supply was suddenly stopped, the total cell population remained constant, but the viable cells decreased to a constant low level.

Hydrogen-ion Concentration of Culture Media

Culture media are adjusted to different degrees of acidity or alkalinity, depending upon the requirements of the organisms under cultivation. Some organism grow best in acid media; others grow best in alkaline media; still others prefer media which are approximately neutral in reaction. This last group includes the great majority of bacteria that have been isolated and studied. Therefore, it is necessary to adjust the reaction of media to the requirements of the organisms being studied.

Two methods are employed for adjusting the reaction of culture media:

1. the determination of the actual numbers of free hydrogen ions and
2. the determination of the net amount of acid or base-binding groups present. The former is spoken of as the hydrogen-ion (H^+) concentration; the latter as the titratable acidity or alkalinity.

The hydrogen-ion concentration may be determined either colourimetrically or electrometrically. The titratable acidity is determined by titration of a known volume of medium with a standard solution of NaOH to the predetermined end point as shown by a glass electrode or by the colour of a suitable indicator. Both method serve very useful purposes in bacteriology. The adjustment of media is more accurately carried out by the hydrogen-ion method. The titratable acidity determination is of great value in learning the buffer content of the medium, i.e., its ability to resist changes in reaction on the addition of acid or alkali.

Measuring the Concentration of Hydrogen (H+) Ions : Pure water is neutral in reaction because it ionizes into equal numbers of hydrogen and hydroxyl ions.

$$H_2O \rightleftharpoons H^+ + OH^-$$

However, it is dissociated to an extremely small extent.

On liter of 1*N* HCl contains approximately 1 gm. of hydrogen (H^+) ions. One liter of pure water contains approximately 0.0000001 gm. of hydrogen ions. This may be written 10^{-7} gm. per liter. For each H^+ ion, there is a corresponding and neutralizing OH^- ion.

According to the law of mass action,

$$\frac{(H^+)(OH^-)}{HOH} = K$$

Since the concentration of undissociated water is very great in comparison to the concentration of free H^+ and OH^- ions, the equation may be written

$$(H^+)(OH^-) = K$$

The numbers of H^+ and OH^- ions being equal, each must have a concentration of 1×10^{-7}. The product of the concentrations of hydrogen and hydroxyl ions is equal to 1×10^{-14}. The equation now becomes

$$(H^+)(OH^-) = 1 \times 10^{-14}$$

Pure water, which has a hydrogen-ion concentration of 1×10^{-7}, is neutral in reaction. Since the product of the H and OH ion concentrations is constant at 1×10^{-14}, when the H^+ increases, the OH^- decreases. The sum of the two always equals 1×10^{-14}. If the hydrogen-ion concentration of a solution is smaller than 1×10^{-7}, it will be alkaline in reaction; if larger than 1×10^{-7}, it will have an acid reaction.

The term pH may be defined as the logarithm of the reciprocal of the hydrogen-ion concentration. For convenience, only the exponent is used in expressing pH. If a solution has a

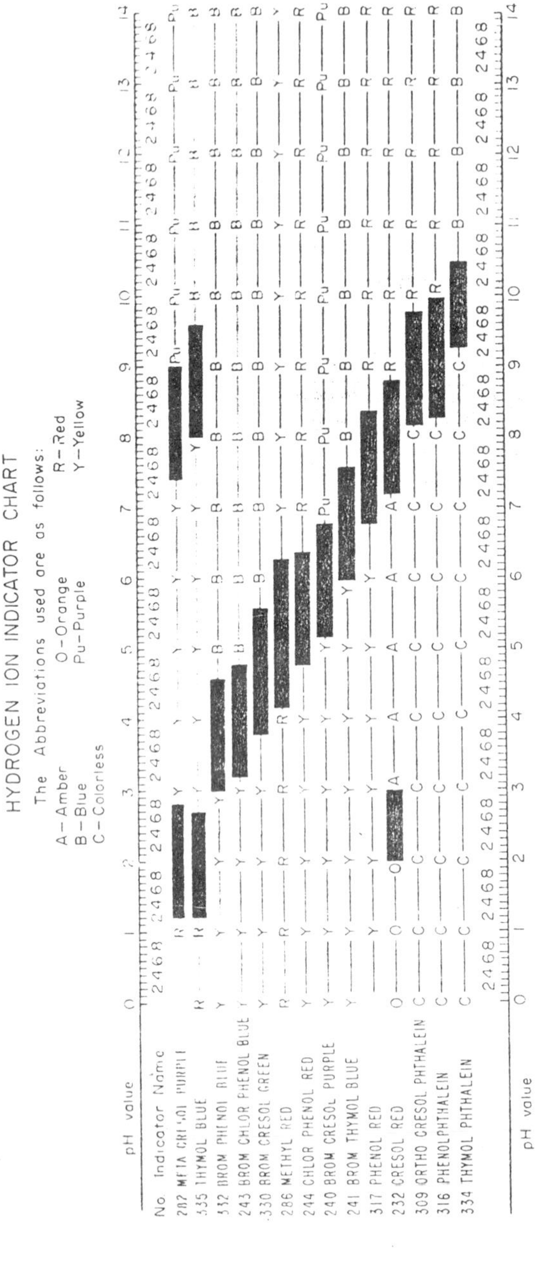

HYDROGEN ION INDICATOR CHART

The Abbreviations used are as follows:

A – Amber
B – Blue
C – Colorless
O – Orange
Pu – Purple
R – Red
Y – Yellow

No.	Indicator Name	0	1	2	3	4	5	6	7	8	9	10	11	12	13	14
282	META CRESOL PURPLE		R		Y	Y	Y	Y	Y		Pu	Pu	Pu	Pu	Pu	Pu
335	THYMOL BLUE	R	R		Y	Y	Y	Y	Y	Y		B	B	B	B	B
332	BROM PHENOL BLUE	Y	Y	Y	Y		B	B	B	B	B	B	B	B	B	B
243	BROM CHLOR PHENOL BLUE	Y	Y	Y	Y		B	B	B	B	B	B	B	B	B	B
330	BROM CRESOL GREEN	Y	Y	Y	Y			B	B	B	B	B	B	B	B	B
286	METHYL RED	R	R	R	R	R			Y	Y	Y	Y	Y	Y	Y	Y
244	CHLOR PHENOL RED	Y	Y	Y	Y	Y			R	R	R	R	R	R	R	R
240	BROM CRESOL PURPLE	Y	Y	Y	Y	Y	Y		Pu	Pu	Pu	Pu	Pu	Pu	Pu	Pu
241	BROM THYMOL BLUE	Y	Y	Y	Y	Y	Y	Y		B	B	B	B	B	B	B
317	PHENOL RED		Y	Y	Y	Y	Y	Y			R	R	R	R	R	R
232	CRESOL RED	O	O	O	A	A	A	A	A		R	R	R	R	R	R
309	ORTHO CRESOL PHTHALEIN	C	C	C	C	C	C	C	C	C		R	R	R	R	R
316	PHENOLPHTHALEIN	C	C	C	C	C	C	C	C	C		R	R	R	R	R
334	THYMOL PHTHALEIN	C	C	C	C	C	C	C	C	C	C		B	B	B	B

pH value: 0–14 (subdivisions 2 4 6 8)

Fig. 7.2 Selector chart for widely used pH indicators.

pH less than 7, it is acid in reaction; if greater than 7, it is alkaline. Most organisms grow best in culture media having a pH of about 7.0.

The relation of pH to H^+ is shown in Table 7.3.

Table— 7.3 The Relation of pH to H^+

	pH	*Normality in terms of hydrogen ions*	*Normality in terms of hydroxyl ions*
Acid	0	1	10^{-14}
	1	10^{-1}	10^{-13}
	2	10^{-2}	10^{-12}
	3	10^{-3}	10^{-11}
	4	10^{-4}	10^{-10}
	5	10^{-5}	10^{-9}
	6	10^{-6}	10^{-8}
Neutral Point	7	10^{-7}	10^{-7}
Alkaline	8	10^{-8}	10^{-6}
	9	10^{-9}	10^{-5}
	10	10^{-10}	10^{-4}
	11	10^{-11}	10^{-3}
	12	10^{-12}	10^{-2}
	13	10^{-13}	10^{-1}
	14	10^{-14}	1

Colourimetric Method : The determination of the hydrogen-ion concentration by the colourimetric method depends upon the colour changes produced in certain weakly acid or weakly basic dyes by varying the reaction of the medium. Such dyes are called indicators. An indicator changes in colour within a short distance each side of that point in the pH scale at which it is 50 percent dissociated. At that point, one-half of the dye is undissociated and the other half is dissociated. The pH at which this occurs is denoted by the symbol pK.

A short distance each side of the pK point gives a zone which is referred to as the sensitive range of the indicator. Every shade of colour of the indicator in this sensitive range corresponds to a definite pH value so that by comparing the shade of the indicator with standards of known reaction, the hydrogen-ion concentration of a solution may be determined.

Indicators can be selected displaying a certain amount of overlapping in their sensitive ranges to cover the scale from pH 1.00 to 10.0

A selector chart for widely used pH indicators is given in Fig. 7.2.

Potentiometric Method : Most pH determinations by the electrometric method are now made with the glass electrode.

Under suitable conditions, a thin glass membrane separating two solutions of different pH values exhibits an electrical potential that is proportional to the difference in the pH of the solutions. For the construction of the glass electrode, a special type fo glass is employed.

The potentiometric method is more accurate than the colourimetric procedure. Also, it can be employed for the determination of the pH value of highly coloured solution which cannot be satisfactorily tested by the use of indicators.

Buffers : The salts fo weak acids have the power of preventing pronounced changes in the reactions of solutions on the addition of relatively large amounts of strong acids or alkalies. Substances which possess the power of resisting changes in acidity or alkalinity are spoken of as buffers.

The titration curve of a weak acid is S-shaped. Each end of the curve has a steep slope and the central portion has a gentle slope. This central, almost horizontal portion of the curve expresses the buffer action of the system or its ability to resist pronounced changes in pH on the addition of acidic or basic substances.

The addition of 1 ml. of *N*/10 hydrochloric acid to 1 liter of neutral distilled water (pH 7.0) gives a solution having a pH of about 4.0. The addition of 1 ml. of *N*/10 sodium hydroxide to 1 liter of neutral distilled water gives a solution having a pH of about 10.0. The addition of the same amount of acid or alkali to 1 liter of distilled water in which are dissolved a few grams of sodium phosphate produces very little change in the reaction. Sodium phosphate is classed as a buffer. This may be shown in the following reactions:

The addition of a strong acid:

$$Na_2HPO_4 + HCl \rightarrow NaH_2PO_4 + NaCl$$

The strong acid (HCl) reacts with the weak alkali (Na_2HPO_4) to give the weak acid (NaH_2PO_4) and sodium chloride. In other words, the strong HCl is replaced by the weak acid phosphate, resulting in a relatively small change in the final hydrogen-ion concentration.

$$NaH_2PO_4 + NaOH \rightarrow Na_2HPO_4 + H_2O$$

The strong alkali (NaOH) reacts with the weak acid (NaH_2PO_4) to give the weak alkali (Na_2HPO_4) and water. The strong NaOH is replaced by the weak basic phosphate, resulting in a relatively small change in the final hydrogen-ion concentration.

The important salts commonly added to nutrient media for their buffering action include phosphates and carbonates. These compounds are particularly valuable because they are relatively non-toxic.

Bacteriological peptones contain such substances as proteoses, peptones, peptides, and amino acids, all of which are buffers. These possess both acidic and basic properties, i.e., have the power of uniting with both bases and acids. Therefore, all culture media containing peptone are well buffered, the degree of buffering being dependent upon the amount of peptone added.

Buffers are special importance in carbohydrate media that are vigorously fermented. In the various fermentations, organic acids are produced. The acidity soon builds up to a concentration that prevents further multiplication of the organisms. In the presence of a buffer, this takes place usually in 24 to 48 hr. In the absence of a buffer, the activity of the organisms may cease after a few hours. A good culture medium, besides containing the necessary nutrients, should also be well buffered.

Buffer Standards : Clark and Lubs (1917) proposed a series of buffer solutions covering the range from pH 1.2 to pH 10.0 at intervals of 0.2 pH. By selection of the proper indicators (Fig. 7.2), these buffer solutions may be used as standards for the adjustment of the reaction of culture media. For this purpose, comparable concentrations of indicator must be used in both the buffer standards and the medium under adjustment.

8

Structure and Classification of Animal Viruses

Viruses are perhaps the simpliest form of life known—simple, at least, when compared structurally with procaryotic and eucaryotic cells. They are able to exist and replicate despite their simple structure because, like several other groups of organisms previously discussed, they are obligate intracellular parasites. But here the comparison of the viruses with rickettsiae or chlamydiae ends. Viruses lack components essential for their own replication and must depend entirely upon the host cell to provide these missing factors.

For example, biologic syntheses require energy, and this energy is provided by ATP in the form of high-energy phosphate bonds, broken to release energy for biosynthetic reactions. Viruses require ATP for a number of very important processes, foremost of which is the replication of their own genetic information. Because viruses lack ATP-generating systems, they rely solely on such a system provided by the host cell.

A virus carries its own genetic information in the form of either RNA or DNA. Virus replication requires expression of this information via the classical pathways of transcription and translation of viral messenger RNAs. However, viruses do not contain all, or, for that matter, most, of the structural elements required for protein synthesis, that is, ribosomes, factors, and tRNAs. Therefore, viruses must utilize ways of subverting normal cellular mechanisms and processes for their own replication. The extent to which they rely on cellular factors for replication depends in large part upon the complexity of their own genetic structure. For example, the large and complex herpesviruses encode many functions that simply replace the host's functions, whereas retroviruses, which encode only three proteins, must rely almost totally on host functions for replication.

Another characteristic peculiar to viruses is that, unlike all other forms of life, viruses contain only one type of nucleic acid; in some cases, it is RNA, and, in others, it is DNA, but never both. Minor exceptions to this statement can be seen in the very complex, DNA-containing poxviruses that apparently contain very small amounts of their own RNA and in the RNA-containing retroviruses, which may contain trace amounts of DNA. However, in both cases, these trace amounts' of "contaminating" nucleic acid are probably not necessary to the virus. The type of nucleic acid contained in a virion greatly influences the nature of its replicative pathway and, hence, is a very useful consideration when classifying viruses.

Sizes of Viruses

Viruses vary considerably in size, but, in general, they are well below the limit of visibility of the light microscope. In early literature, they were often referred to as filterable agents, because they would pass through filters that would normally retain bacteria. In fact, early errors concerning the etiology of a disease have occurred when a virus was considered to be the causative agent merely because the infectious agent was filterable.

Virus sizes range from 20 nm to 250 nm (nm = 10^{-9} meter, or one thousandth of a μm), although, as will be seen, such measurements can be misleading for helical-shaped viruses such as the rhabdoviruses. Classically, three basic techniques are used to determine virus size: (1) filtration through graded collodian membranes for which the size of the pores in each membrane is known: (2) high-speed centrifugation (greater than 1,00,000 times gravity), which provides data for the calculation of virus size based on the rate at which a particle of known shape will travel toward the bottom of the centrifuge tube at a known centrifugal force: and (3) direct observation using an electron microscope, often including objects of known size in the preparation for comparison. Modern day instrumentation makes the latter the method of choice today.

CHEMICAL COMPOSITION AND STRUCTURE OF VIRUSES

Viruses viewed in their simplest form are nothing more than nucleic acid (either DNA or RNA) surrounded by a protein overcoat called a *capsid*. This simple characteristic becomes more complex because may viruses contain additional structural proteins and enzymes. Also, some may have carbohydrates bound to the surface proteins (glycoproteins), and others may be enclosed in a membrane that is composed of virally coded proteins and host-cell lipids. Our discussion of virus structure begins with the characterization, function, and structure of a basic viral component, the viral capsid.

Viral Capsids

Nature has designed the viral capsid to perform several functions. The viral capsid protects the enclosed nucleic acid from both physical destruction and enzymatic hydrolysis by the extracellular milieu. It also provides binding sites that enable the virus to attach to specific receptor sites on the host cell. Electron microscopy has revealed that, with the exception of the relatively complex poxviruses, all animal viruses are either isometric or helical in shape. Virtually all DNA containing animal viruses are icosahedrons, an exception being brick-shaped poxviruses; RNA-containing animal viruses may exist as icosahedrons or as helical viruses.

Isometric Viruses : Isometric viruses appear to be in the shape of an icosahedron. Each particle has 20 facets, each an equilateral triangle. These facets come together to form 12 vertices of 5 facets each. The completed capsid is made up of repeating morphologic units called *capsomers*. These capsomers are visible by electron microscopy as small protein structures. For some viruses, each capsomer my be composed of only a single repeating polypeptide, whereas in other viruses, each capsomer may consist of several different polypeptide molecules held together by non-covalent bonds to form the completed capsomer. As shown in the figure 8.1 the capsomer at each of the 12 vertices is surrounded by only 5 other capsomers, whereas all other capsomers are adjacent to 6 other capsomers. The capsomers at the vertices are called pentons, or pentomers, and the remaining capsomers are referred to as hexons, or hexomers.

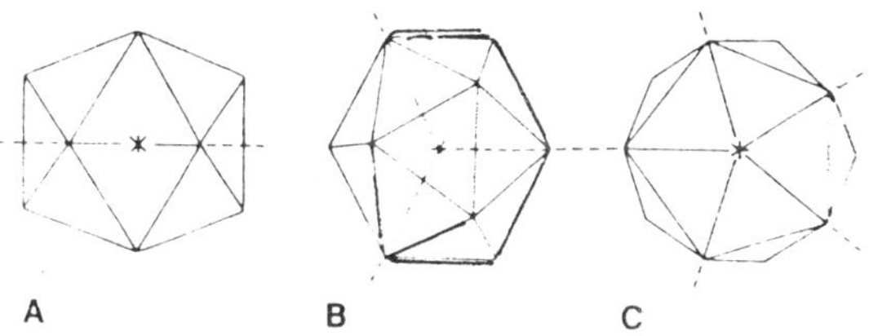

Fig. 8.1 Features of icosahedral symmetry.

A special property of an icosahedral structure is its multiple axes of symmetry. It is possible to discern three different types of symmetry as one rotates the icosahedron, through the edges, the faces, and the vertices, resulting in what is referred to as 5:3:2 symmetry.

One means of classifying virtuses is based on the number of capsomers present in the viral capsid. This number can be calculated for many isometric virions by using the formula $N = 10(n - 1)^2 + 2$, where n is the number of capsomers on one side of each equilateral triangle. Examples that fit this formula would include herpesvirus, with 5 capsomers on the side of each triangular surface and thus a total of 162 capsomers making up the intact capsid, and adenovirus, with 6 capsomers per side and a total of 252 capsomers per intact virion.

Helical Viruses : In the case of helical viruses, the viral nucleic acid is closely associated with the protein capsid, forming a coil-shaped nucleocapsid that becomes enclosed in a membrane as it buds from the host cell. The exact nature of the nucleic acid interaction with the capsid is not well understood, but the protein protects the RNA from enzymatic degradation by nucleases while still allowing the RNA to be transcribed from the intact nucleoprotein. Electron micrographs of rhabdovirus virions in which the striation of nucleic acid can be seen wound in a helical pattern in the interior of the particles. With the exception of the bullet-shaped rhabdoviruses, other helical animal viruses exist as spherical virions containing a helical-shaped nucleocapsid surrounded by a membrane.

Viral Nucleic Acid

Each family of viruses possesses a nucleic acid characteristic for that group, which, along with the properties of the virion capsid, is used for the classification of the virus. Viruses may possess double-stranded DNA, (ds RNA), single- stranded DNA (ss DNA), ds RNA, or ss RNA. Furthermore, the amount of nucleic acid present may vary from enough to code for only three or four proteins (e.g., parvoviruses, picornaviruses) to that seen in the herpesviruses and poxviruses, which can code for hundreds of proteins.

Structure of Viral DNA

Differences in viral DNA structure dictate that viruses evolve different strategies for replicating their genetic information. As will be seen later, the elucidation of these mechanisms by the molecular biologist has yielded invaluable insights into the mechanisms of cellular DNA replication and has provided model systems for the analysis of DNA replication.

Supercoiling : The DNA of most animal viruses exists as a linear molecule, either ds DNA or ss DNA, or a covalently closed, supercoiled circle. Supercoiled DNA results from extra turns in the helical structure of ds DNA, which are introduced by an enzyme called DNA gyrase. Supercoiled DNA can be converted into a simple circular structure by enzymatically breaking (nicking) one strand of the ds DNA. This relieves the supercoiled twist, permitting the relaxation of the molecule to form a "relaxed" circular molecule. Supercoiling occurs in all types of DNA, mitochondrial DNA, chloroplast DNA, and bacterial plasmid DNA as well as viral DNA. It undoubtedly provides an important organizational function for certain DNA molecules by permitting a higher-order structuring of the DNA.

Terminal Repetitions : Molecular biologists have long appreciated the fact that bacteriophage DNAs frequently have unique sequence arrangements at their ends. Bacteriophage λ contains complementary termini which make it possible for the DNA to form circular structures during the process of DNA replication (**Fig. 8.2**). The termini of the DNA of many viruses, such as adenoviruses and parvoviruses, contain inverted repeat sequences. The terminal sequence on each strand of DNA may be represented as 5′-ABC ......C′B′A′ (where ABC are complementary to A′B′C′). Such sequence organization can be visualized in the electron microscope if the ds DNA is heated and the separated strands are permitted to reanneal. In such a case, part of the DNA would base-pair to reform the original ds DNA, and part would base-pair its 3′ and 5′ ends so as to form a panhandle or lollipop structure smilar to that described for a transposon. Herpesvirus DNA possesses a more complicated organization of repeated DNA sequences. Sequences complementary to the terminally repeated sequences are also found within the viral DNA sequnece, providing an internal repetition.

The existence of terminal redundancy in viral DNA provides a unique structural feature important for the replication of the virus. Such terminal redundancy makes it possible for the same protein to recognize both ends of ds linear DNA. In addition, inverted terminal repetitions in ss DNA may provide a double-stranded region necessary for the initiation of replication of the ss DNA.

Structure of Viral RNA

The RNA in animal viruses exists only as linear double-strand or single-strand molecules. Unlike DNA, however, the RNA within a virion may exist as a segmented genome. For example, reoviruses contain 10 different molecules of ds RNA; rotaviruses contain 11 segments of ds RNA, and influenza viruses possess 8 separate segments of ss RNA. Other RNA viruses containing a segmented genome include the bunyaviruses and the arenaviruses. Retroviruses contain two identical single-strand genomes. Such a complex organization of genetic information requires unique mechanisms to ensure the proper segregation of genetic material and also provides the viruses with unique opportunities for the alteration of the genetic composition of the virus.

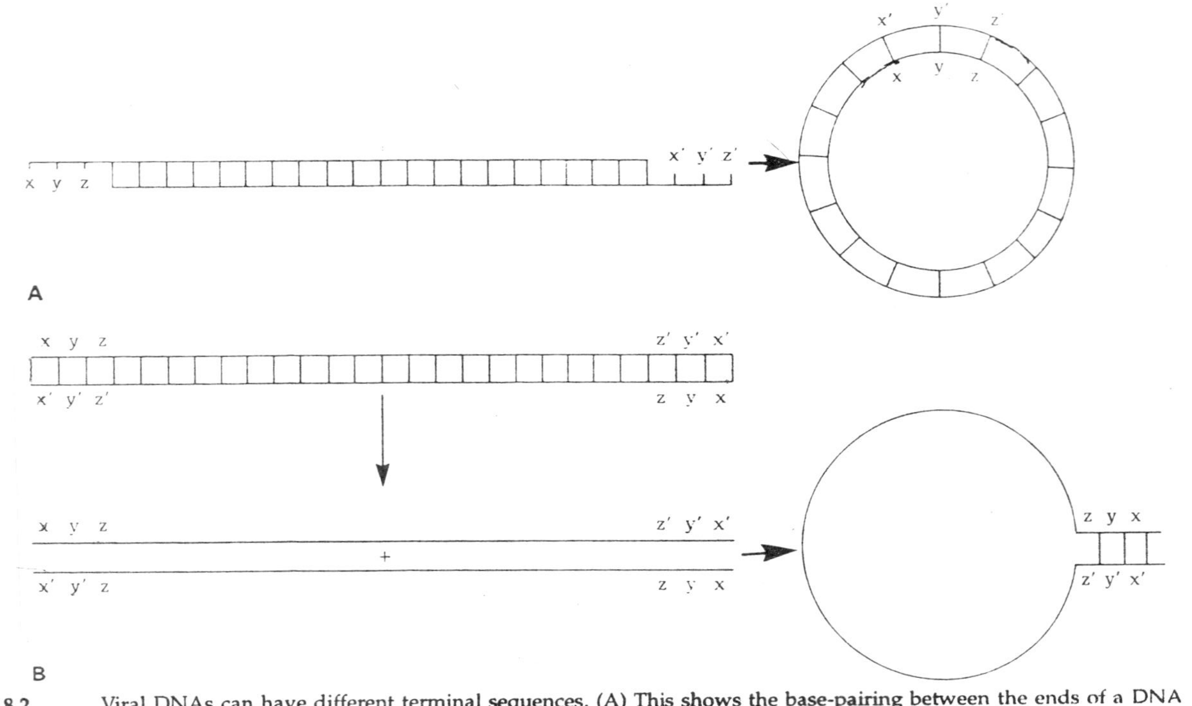

Fig.8.2 Viral DNAs can have different terminal sequences. (A) This shows the base-pairing between the ends of a DNA molecule with cohesive termini. (B) This shows the possible base-pairing between the ends of a DNA molecule containing an inverted repreated sequence.

Plus and Minus Strands of Nucleic Acid

Viruses containing single-stranded nucleic acid package only one of the two strands within the virion. A useful convention has been established to distinguish viruses containing ss RNA of different polarities. Those viruses containing ss RNA, which acts directly as messenger RNA (mRNA), have been designated as plus (+) strand viruses, whereas those that must first replicate their RNA to form a complementary strand (which acts as the mRNA) are designated as minus (–)-strand viruses. As one might expect, the naked RNA of plus-strand viruses, such as poliovirus, can infect an animal cell, directing the synthesis of complete virus particles, because the entire message is carried on the mRNA, which is read by the host-cell enzymes. Naked RNA from minus-strand viruses, however, is not infectious, because it must first be replicated to form the complementary mRNA. Such replication is always catalyzed by an RNA polymerase carried within the virion.

DNA viruses can also exist as plus or minus ss DNA, but ss DNA viruses must be converted to ds DNA in order to be transcribed by RNA polymerase to form mRNA. During transcription of ds DNA, generally only one strand is read, and one can consider that strand to be plus DNA and its complementary strand to be minus DNA. Classification of a virus as either (+) or (–) strand virus is extremely useful because it immediately reveals certain features of the strategy of virus replication.

Viral Envelopes

It has already been caused a virion as consisting of a protein capsid enclosing the viral nucleic acid and, in some cases, one or more viral enzymes. This combination of capsid and nucleic acid is referred to as the nucleocapsid, and, in the case of some animal virus groups, this constitutes the completed virion. A large number of animal viruses, however, contain a capsid surrounded by a lipid envelope acquired during the final stage of replication as they bud through special areas of the host cell membrane. These special areas of host cell membrane are areas in which the host cell proteins have been supplanted by viral-coded polypeptides and virus-specified glycoproteins, although it is likely that the carbohydrate moiety of the glycoproteins is specified by the host cell. The glycoprotein usually occurs as projections or spikes on the outer surface of the envelope. These protrusions have been given the name peplomers. Many of these glycoproteins have special functions, in particular, mediatig the attachment of the virus to host cell receptors to initiate the entrance of the virion into the cell. Some glycoproteins will also attach to receptors on red blood cells, causing these cells to agglutinate.

9

Taxonomy of Viruses

Viruses are separated into families on the basis of the type and form of the nucleic acid genome and the size, shape, substructure, and mode of replication of the virus particle. Within each family, classifications of general and species are based on antigenicity in addition to other properties.

Significant development in classification and nomenclature of viruses are documented in the reports of the International Committee on Taxonomy of Viruses, formerly the International Committee on Nomenclature of Viruses. These reports, published in 1971 (6), 1976 (2), 1979 (3), and 1982 (4), have dealt with viruses of humans, lower animals, insects, plants, and bacteria and have included summaries of the properties of those groups of viruses as related to their taxonomic placement. It seems probable that most of the major groups of viruses have been recognized, particularly with regard to those infecting humans and the vertebrate animals of direct importance to humans. Many of them now have been officially placed in families, genera, and species; within some families, subfamilies or subgenera or both also have been established. Although the focus of this chapter is on these families, progress also has been made with respect to the taxonomy of the viruses of other host groups. For more detailed discussion of virus taxonomy, references 1 through 6 are recommended to the reader.

In Table 9.1, properties of the major families of RNA-containing viruses of humans and other vertebrate animals are summarized; in Table 9.2, vertebrate viruses with a DNA genome are similarly treated. In Table 9.3, these families are listed along with the genera and the individual members that may be of special concern for the viral diagnostic laboratory. Where subfamilies have been designated, these also are included if they represent agents having direct or indirect bearing upon human diseases. Since most of these virus agents are dealt with more fully in the other chapters of this volume, the brief text which follows is confined to explanation and commentary on some examples from the tables, together with some notes about recent developments.

Picornaviridae : Of particular importance is the recent classification of hepatitis A virus as enterovirus 72, within the family Picornaviridae. This virus has been shown to have the physicochemical properties of a member of the genus *Enterovirus*. These properties include a non-enveloped icosahedral (cubic) virion about 27 nm in diameter, a buoyant density in Cs CI of ca. 1.33 to 1.34 g/ cm^3, and four major polypeptides with molecular weights of about 33,000, 27,000, 23,000 and 6,000. The genome consists of a single piece of single-stranded RNA of molecular weight ca. 2.5×10^6. Like the other enteroviruses, hepatitis A virus is stable to acid pH and resistant to ether. In its resistance to thermal inactivation this serotype differs somewhat from other enteroviruses. In comparative studies, 50% of the particles in a poliovirus type 2 preparation disintegrate during heating at pH 7 for 10 min at 43°C, whereas under the same conditions 61°C is required to produce disintegration of 50% of the hepatitis A virus (enterovirus 72) particles. However, enterovirus 72, like all other enteroviruses, is stabilized by $MgCl_2$ against thermal inactivation. In the presence of 1 M $MgCl_2$, the temperatures required to produce 50% destruction of the particles are 61°C for poliovirus type 2 and 81°C for enterovirus 72. Stabilization of infectivity by $Mgcl_2$ is a well-established characteristic of enteroviruses. Enterovirus 72 shows unusual thermal resistance without $MgCl_2$ (loss of only

1.3 logs of infectivity after heating at 65°C for 10 min), but in the presence of $MgCl_2$ it is even more resistant. There is no reduction after 20 min at 65°C, and only 1.5 log units of infectivity is lost after 10 min at 80°C.

Reoviridae : For all of the virus families listed in Table 9.1, the RNA genome is single stranded except in the case of the family Reoviridae, whose RNA is double stranded. The genus *Reovirus* differs somewhat from the other genera in its possession of an outer protein shell and the larger molecular weight of its genome (15×10^6 versus 12×10^6). There are three serotypes within the *Reovirus* genus that infect humans, monkeys, dogs, and cattle; in addition, at least five avian reoviruses are known. In the genus *Orbivirus*, important human pathogens include Colourado tick fever virus and the Kemerovo viruses; other important orbiviruses are the bluetongue viruses of sheep and the viruses of African horse sickness. The most recently established genus is *Rotavirus*, which includes several viruses that infect humans as well as viruses of many other mammalian species, typified by simian virus SA-11 and Nebraska calf diarrhea virus. The human rotaviruses are increasingly being recognized as the cause of a large share of the serious episodes of non-bacterial infantile diarrhea. Rotavirus gastroenteritis is one of the most common childhood illnesses throughout the world and is a leading cause of infant deaths in developing countries. These viruses also infect adults, particularly those in close contact with infants and children, but infected adults usually experience no symptoms or may have only minor illness.

Caliciviridae : Other recent additions to the taxonomic roll of RNA- containing viruses are the Caliciviridae and the Bunyaviridae. The Caliciviridae include a number of viruses of pigs, cats, and sea lions and may include agents that infect humans. Calicivirus-like particles have been observed in human feces in association with gastroenteric disease; preliminary results have failed to show relationship to the feline calicivirus. The possible relationship of these agents to the virus of Norwalk gastroenteritis also remains to be resolved. The Norwalk virus, a widespread human agent causing acute epidemic gastroenteritis, has a virion protein structure similar to that of

Table 9.1— Current classification of RNA-containing viruses of vertebrates

Characteristic	Classification										
Nucleic acid core	RNA										
Capsid smmetry	Cubic				Heical					Uncertain	
Virion: naked or enveloped	Naked			Enveloped	Enveloped					Enveloped	
Site of capsid Assemly	Cytoplasm			Cytoplasm	Cytoplasm					Cytoplasm	
Site of nucleocapsid envelopment				Surface membrane[a]	Surface membrane			Intracytoplasmic membranes		Surface membrane	
Reaction to ether treatment	Resistant			Sensitive	Sensitive			Sensitive		Sensitive	
Number of Capsomeres	32	32	32	32							
Diameter of helix (nm)					9-15	18[b]	18	11-13	10-12		
Diameter of varion (nm)[c]	24-30	35-39	60-80	40-70[a]	80-120	150-300	60 x 80	80-130	80-110	about 100	50-300

(Contd...)

Table 9.1 (Contd...)

Mol wt of nucleic acid (× 10)[6]	2.3-2.8	2.6	12–15	3–4	4–5	5–8	3.5-4.6	5–6	6–7	6–7	3–5
Virus family	Picornaviridae	Calciviridae	Reoviridae	Togaviridae	Orthomyxoviridae	Paramyxoviridae	Rhabdoviridae	Coronaviride	Bunyaviridae	Retroviridae	Arenaviridae

[a] All of the properties shwo for Togaviridae apply to the genera *Alphavirus* and *Rubivirus*; however, for the genera *Pestivirus* and *Flavivirus*, envelopment takes place at intracytoplasmic membranes, the number of capsomeres is not clearly established, and the virions are somewhat smaller.

[b] All of the properties shown for Paramyxoviridae apply to the genera *Paramyxovirus* and *Morbillivirus*; however, for members of the genus *Pneumovirus*, the diameter of the helix is 12 to 15 nm.

[c] Diameter of diameter × length.

the caliciviruses; it also resembles caliciviruses in several other characteristics. Because these agents have not yet been successfully adapted to tissue culture, it has been difficult to study their properties.

Bunyaviridae : The Bunyaviridae form a family of more than 200 viruses, at least 145 them belonging to the Bunyamwera supergroup of serologically interrelated arboviruses. With the taxonomic placement of this large group, the vast majority of the viruses of the classical arbovirus groupings, initially based on ecological properties and subdivided by serological interrelationships, have been assigned to families on the basis of biophysical and biochemical characteristics. Human illness caused by Hantaan virus has been recognized in the Far East as Korean "hemorrhagic fever, and a variant is known in Scandinavian and eastern European countries as epidemic nephropathy. The illness has been variously named "hemorrhagic fever with renal syndrome" or "muroid virus nephropathy." Recent studies show the causative agent to be a member of the Bunyaviridae. The virus has a labile membrane and a tripartite single-stranded RNA genome. The most common natural hosts are mice (in Korea) and voles (in Europe). There have been several instances of infection of staff members handling laboratory rats infected with the virus, both in the Far East and more recently in Europe. In Belgium there also have been sporadic cases with no apparent link to an outbreak or to each other.

Retroviridae : The family Retroviridae has been divided into subfamilies (see Table 9.3). The best-known retroviruses belong to the subfamily Oncovirinae, the RNA tumour virus group, which has been the focus of special interest because its members, long recognized as causing leukemia and sarcoma in animals, serve as valuable animal models of oncogenic viruses.

The Retroviridae are enveloped viruses whose genome contains a single-stranded RNA of the same polarity as viral mRNA. The virion contains a reverse transcriptase enzyme. Replication proceeds off an integrated "provirus" DNA copy in infected cells. Study of retroviruses, and oncoviruses in particular, has permitted the identification of cellular

Table— 9.2 Current classification of DNA-containing viruses of vertebrates

Characteristic	*Classification*						
Nucleic acid core	DNA	DNA	DNA	DNA	DNA	DNA	DNA
Capsid symmetry	Cubic	Cubic	Cubic	Cubic	Cubic	Cubic	Complex
Virion: naked or enveloped	Naked	Naked	Naked	Enveloped	Enveloped	Enveloped	Complex coats
Site of capsid assembly[a]	Nucleus	Nucleus	Nucleus	Nucleus	Nucleus	Cytoplasm	Cytoplasm
Site of nucleocapsid envelopment				Cytoplams	Nuclear membrane	Cytoplasmic membrane	
Reaction to ether (or other lipid solvents)	Resistant	Resistant	Resistant		Sensitive	Sensitive	Resistant
Number of capsomeres	32	72	252		162	1,500	
Diameter of virion (nm)[b]	18–26	45–55	70–90	40–50	100[c]	130–300	230 x 300
Mol wt of nucleic acid (x 10^6)	1.5–2.0	3.0–5.0	20–30	2.1	80–150	100–250	160
Virus family	Parvoviridae	Papovaviridae	Adenoviridae	Hepadnaviridae	Herpesviridae	Iridoviridae	Poxviridae

a For the DNA- containing viruses whose capsid assembly takes place in the nucleus, a pahse of replication occurs in the cytoplasm, as evidenced by the detection of viral mRNA associated with polyribosomes.

b Diameter or diameter x length.

c The naked virus is 100nm in diameter; however, the eveloped virions range up to 150 nm in diameter.

"oncogenes." Normal cells of several animal species contain integrated copies of the genes of the endogenous oncovirus. The oncovirus genes may not be expressed but can be activated by physical and chemical agents, by superinfection with other oncoviruses, and even by herpesviruses, Recently, members of the subfamily Oncovirinae have been shown to cause human disease. These include the agents generally known as the human T-cell leukemia viruses and the agent known as the lymphadenopathy-associated virus. There are at least three types of human T-cell leukemia viruses, and type 3 apparently is identical with the lymphadenopathy-associated virus; this serotype seems to be the etiological agent of AIDS (acquired immune deficiency syndrome).

Parvoviridae : All of the virus families shown in Table 9.2 have their DNA genome in double-stranded form except the Parvoviridae, whose DNA is single stranded within the virion. As indicated in Table 9.2, members of Parvoviridae are very small viruses. The molecular weight of the nucleic acid in the virion is relatively very low, 1.5×10^6 to 2.0×10^6 (as compared, for example, with 160×10^6 for the DNA of poxviruses). Some members display resistance to high temperatures (60°C, 30 min).

The family Parvoviridae encompasses viruses of numerous species of vertebrates, including humans. Two members of the genus *Parvovirus,* the members of which are able to replicate independently, have been found to be associated with disease problems of human beings. Parvovirus B19 has been shown to cause a transient shutdown of erythrocyte production by killing the late erythroid progenitor cells. This shutdown presents particular problems for individuals already suffering from hemolytic anemias such as sickle cell anemia, causing aplastic crises. A virus named RA-1, which is associated with rheumatoid arthritis, is another newly identified member of the genus.

A host range mutant of feline panleukopenia parvovirus, known as canine parvovirus, induces acute enteritis with leukopenia in young and adult dogs as well as myocarditis in puppies. Infections with this virus have reached enzootic proportions around the world.

Several serotypes of adeno-associated viruses, belonging to the *Dependovirus* genus, are known to infect humans, but they have not been shown to be associated with any human disease. Members of this genus cannot multiply in the absence of a replicating adenovirus which serves as a "helper virus." The singlestranded DNA is present within the virion as either plus or minus complementary strands in separate particles. Upon extraction, the plus and minus DNA strands unite to form a double-stranded helix.

Papovaviridae : Members of the family Papovaviridae have DNA in double-stranded, circular form. The human representatives, are the papilloma or wart viruses and the JC and BK viruses; these latter were isolated, respectively, from the brain tissue of patients with progressive multifocal leukoencephalopathy and from the urine of immunosuppressed renal transplant recipients. In addition, several isolates which appear to be identical to simian virus 40 of monkeys have also been isolated from patients with progressive multifocal leukoencephalopathy. Papovaviruses produce latent and chronic infections in their natural hosts. Many of them produce tumours, particularly in experimentally infected rodents, thus serving as models for studying viral carcinogenesis. The viral DNA integrates into cellular chromosomes of transformed cells.

When simian virus 40 and adenoviruses replicate together, they may interact to form "hybrid" virus particles, in which a defective simian virus 40 genome is covalently linked to adenovirus DNA and is carried within an adenovirus capsid.

Hepadnaviridae : Ample evidence has accumulated for the formation of a new virus family. The name, Hepadnaviridae, reflects the DNA-containing genomes of its members and their replication within hepatocytes. These viruses have a circular DNA genome that is double stranded except for a region of variable length that is single stranded. In the presence of appropriate substrates, DNA polymerase within the virion can complete the single-stranded region to its full length of 3,200 nucleotides.

Table— 9.3 **Members of virus families, with emphasis on viruses that infect humans**

Family	*Genus*	*Common species*	*No. of members*
Picornaviridae	*Enterovirus*	Polioviruses	3
		Coxsackieviruses, group A	23
		Coxsackievirues, group B	6
		Echoviruses	31
		Enteroviruses 68 through 71	4
		Enterovirus 72 (hepatitis A virus)	1
		Viruses of other vertebrates	>34
	Cardiovirus	Encephalomyocarditis virus and mengovirus, mouse encephal-omyelitis virus	3
	Rhinovirus	Virus types infecting humans	>115
		Viruses of cattle	2
	Aphthovirus	Foot-and-mouth disease viruses of cattle and other colven-hoofed animals	7
Caliciviridae	*Calicivirus*	Vesicular exanthema of swine virus	13
		Viruses of cats, sea lions	Many
		(Possible member: Norwalk gastroenteritis virus of humans)	?
Reoviridae	*Reovirus*	Viruses of humans, monkeys, and lower vertebrates	3
		Viruses of birds	>5
	Orbivirus	Seventeen subgroups, including Colorado tick fever and Kemerovo viruses of humans and also bluetongue virus of sheep and African horse sickness virus	>90
	Rotavirus	Humans rotaviruses	>4
		Rotaviruses of many mammals, including SA- 11 virus of monkeys and Nebraska calf diarrhea virus	Many
Togaviridae	*Alphavirus* (arbovirus group A)	Sindbis virus and many other mosquito-borne viruses, including the viruses of eastern equine, Venezuelan, and western equine encephalitis and Semliki Forest virus.	23
	Flavivirus (arbovirus group B)	Yellow fever virus and other mosquito-borne viruses, including the viruses of dengue, of Japanese, Murray Valley, and St. Louis encephalitis, and of West nile fever	26

(Contd...)

Table 9.3 (Contd...)

Family[a]	*Genus*	*Common species*	*No. of members*
		Tick-borne viruses, including the viruses Kyasanur Forest disease, Omsk hemorrhagic fever, European and Far Eastern tick-borne encephalitis of humans, and louping ill of sheep.	11
		Viruses whose vectors are unknown	17
	Rubivirus	Rubella virus	1
	Pestivirus	Viruses of cattle and pigs	>3
Orthomyxoviridae	*Influenzavirus*	Influenza virus type A	Many
		Influenza virus type B	Several
	(Probably a separate genus)	Influenza virus type C	1
Paramyxoviridae	*Paramyxovirus*	Human parainfluenza virus, including Sendai virus	4
		Mumps virus	1
		Newcastle disease virus of fowl and viruses of other diseases of birds and mammals	>6
	Morbillivirus	Measles virus	1
		Rinderpest virus of cattle	1
		Distemper virus of dogs	1
		Peste-des-petits-ruminants virus of sheep and goats	1
	Pneumovirus	Human respiratory syncytial virus	1
		Respiratory disease viruses of cattle and of mice	?
Rhabdoviridae	*Vesiculovirus*	Vesicular stomatitis virus of , horses, cattle and pigs	Several
	Lyssavirus	Rabies virus	1
		Lagos bat virus and others	>5
Coronaviridae	*Coronavirus*	Human coronavirus	1
		Mouse hepatitis virus, infectious bronchitis virus of fowl, and other agents infecting pigs and other vertebrates	>4
Bunyaviridae	*Bunyavirus*	Bunyamwera virus California encephalitis viruses	>145

(Contd...)

Table 9.3 (Contd...)

Family	*Genus*	*Conunon species*	*No. of members*
		LaCrosse virus, other serologically cross-related groups, and several ungrouped viruses	
	Phlebovirus	Sandfly fever viruses Other viruses of humans and animals, including Rift Valley fever virus of sheep and other ruminants, which may cause human disease	>30
	Nairovirus	Crimean-Congo hemorrhagic fever Viruses of five other serogroups, including the virus of Nairobi sheep disease	>27
	Uukuvirus	Unkuniemi virus and six other agents, all belonging to the same serogroup (infect rodent and ticks)	7
	Hantaanvirus	Hantaan virus of Korean hemorrhagic fever (hemorrhagic fever with renal syndrome)	1
Retroviridae Oncovirinae[a] (RNA tumour virus group)	Type C oncovirus group	Sarcoma and leukemia viruses of mice, cats, cattle, birds, snakes, and primates, including HTLV and LAV[b] of humans	>15
	Type B oncovirus group	Mammary tumour virus of mice (and humans?)	?
	(Proposed genus: type D retrovirus group)	Monkey (mammary tumour?) virus (Mason-pfizer monkey virus)	?
Spumavirinae (foamy virus Group)		Syncytial and foamy viruses of humans, monkeys, cattle, and cats	>4
Lentivirinae (maedivisna virus group: "slow viruses")		Visna virus of sheep, maedi, progressive pneumonia viruses of sheep	?
Arenaviridae	*Arenavirus*	Lymphocytic choriomeningitis virus of mice	1
		Lassa fever virus	1
		Viruses of the Tacaribe complex, including Junin and Machup'o	>8

(Contd...)

Table 9.3 (Contd...)

Family[a]	*Genus*	*Common species*	*No. of members*
		viruses of South American hemorrhagic fevers	
Parvoviridae	Parvovirus	Human parvoviruses: B19, RA-1	>2
		Aleutian mink disease virus and viruses of rodents, pigs, cattle, cats, and dog	Many
	Dependovirus	Adeno-associated virus (adeno-satellite virus): human (types 1-3); monkey (type 4); also of cattle, dogs, birds	>8
Papovaviridae	*Papillomavirus*	Human papilloma (warts) viruses	Many
		Rabbit (Shope) papilloma virus	1
		Papilloma viruses of other mammals	Many
	Polyomavirus	Polyoma virus of mice	1
		JC and BK viruses of humans	2
		Simian virus 40 of rhesus monkey	1
		Lymphotropic virus of African green monkey	1
		Viruses of mouse, rabbit, hamster, and baboon	4
Adenoviridae	*Mastadenovirus*	Human adenoviruses	36
		Viruses of other mammals	>45
	Aviadenovirus	Viruses of birds	>13
Herpesviridae			
Alphaherpes-virinae[a]	*Simplexvirus*	Human herpes simplex virus types 1 and 2	2
		Bovine mammillitis virus	1
		Herpes B virus of monkeys	1
	Poikilovirus	Pseudorabies virus Equine rhinopneumonitis virus	2
	Varicellavirus	Varicella-Zoster virus	1
Betaherpesvirinae	*Cytomega-lovirus*	Human cytomegaloviruses	1
	Muromega-lovirus	Mouse cytomegaloviruses	1

(Contd...)

Table 9.3 (Contd...)

Family[a]	*Genus*	*Common species*	*No. of members*
Gammaherpes-virinae	*Lymphocry-ptovirus*	Epstein-Barr virus	1
	Thetalympho-cryptovirus	Marek's disease herpesvirus of fowl	1
	Rhadinovirus	Herpesvirus saimiri and others	>2
Iridoviridae (icosahedral cytoplasmic deoxyriboviruses)	African swine fever virus group	African swine fever virus	Several
	Ranavirus	Frog virus 3 and other viruses of amphibians	>30
Poxviridae			
Chordopox-virinae[a] (poxviruses of vertebrates)	*Orthopoxvirus*	Vaccinia virus Smallpox virus (variola) Poxviruses of lower animals	1 1 >6
	Parapoxvirus	Orf virus and other viruses of ungulates Virus of milker's nodule	? 1
	Avipoxvirus	Fowlpox virus and other viruses of birds	8
	Capripoxvirus	Viruses of sheep and goats	3
	Leporipoxvirus	Myxoma virus of hares Fibroma viruses of rabbits and squirrels	4
	Suipoxvirus	Swinepox virus	1
Entomopoxvirinae		Poxviruses of insects	>24
Hepadnaviridae	*Hepadnavirus*	Human hepatitis B virus Hepatitis viruses of woodchuck, ground squirrel, and duck	1 >3

a Where subfamilies have been designated, they are listed in this column, indented below the family name (see Retroviridae, Herpesviridae, and Poxviridae).

b HTLV, Human T-cell leukemia viruses; LAV, lumphadenopathy-associated virus.

Hepatitis B virus of humans and three similar viruses found in woodchucks, Beechey ground squirrels, and Pekin ducks share many basic features. All members of the family share antigens as well as similar morphology and behaviour in the infected host. Large amounts of excess viral coat protein are produced in the form of small 22-nm spherical and tubular particles (in the human virus, the antigen is known as hepatitis B surface antigen). The viruses replicate in the liver and are associated with acute and chronic hepatitis. More than 200 million persons are persistent carriers of the human virus and are at very high risk of developing liver cancer. The woodchuck hepatitis B virus also causes liver cancer in its natural host. Fragments of viral DNA may be found in the liver cancer cells of both species.

10

Pathogenesis of Virus Infections

All viruses multiply in the intracellular position, and in the course of this growth disturb the physiologic state of the invaded cell. Often this disorder shows itself as a degeneration, but sometimes as a proliferation. In either instance, secondary inflammation may be present. These changes constitute a pathologic picture characteristic of the virus concerned or of the family of viruses to which it belongs.

Many advances have improved our understanding of the basic pathology of virus infections. Of great importance has been work on pathogenesis, which has made it plain that many viruses spread through the body by the blood stream and may localize in several bodily organs, as well as in the organ most evidently affected on clinical examination.

The new technics of tissue culture have provided an elegant experimental model for the study of virus action on the individual cell. For convenience, histologic changes in tissue cultures induced by viruses have been discussed earlier. These studies have emphasized the fact that individual viruses tend to affect the cell in different ways. Some cause rapid necrosis. Others disturb the invaded cell to a much lesser degree, hile some apear to be carried as latent agents without causing any cytopathology. Of considerable interest has been the finding that many viruses stimulate in tissue cultures the formation of syneytial cells with multiple nuclei. This is seen in the human body also, especially in skin and lymphoid tissue.

The technic of ultramicrotomy has revealed that many inclusions well known to morbid histologist s are in fact intracellular aggregations of virus particles. The fluorescent antibody technic is another means of localizing in the cell the presence of viral antigenic material.

It is being realized, largely from the use of monkey organs as materials for tissue culture, that latent viral infection of man and animals may be quite common. As described in chapter 4, large numbers of viruses have been recovered from the organs and execrations of monkeys. These viruses do not seem often to cause pathologic changes, although they may do so, as for example in monkey herpes (infection with virus B). These observations raise the possibility that in man also viruses may multiply and be excreted without causing any significant pathology in the organs concerned. It has of course been known for many years that herpes simplex virus may cause a lifelong latent infection. Poliovirus and other enteroviruses likwise probably cause a latent infection of the gastrointestinal tract, and the common cold and other viruses latently infect the respiratory tract.

A related problem is the tendency for viruses to persist in the body long after recovery from the acute phase of the illness. One of the classic examples is, of course, Brill's disease, where *Rickettsia prowazekii* becomes active and causes illness many years after apparent recovery from the primary infection. In psittacosis also, the virus has been recovered a long time after the initial infection. The demonstration that viruses may persist in the body in a latent form, without causing widespread pathologic changes, is of obvious interest. There is a direct relationship here to the etiology of malignant disease, and the possibility of a tumour-producing virus being carried in the body, for years and eventually being stimulated to cause clinical disease by factors unknown.

Pathogenesis

A description of pathogenesis includes the mode of entry of a virus into the body, local proliferation at the site of entry, dissemination of virus in the body, localization in the "target" organ bearing the brunt of the attack, and subsidiary localization in other organs.

In general, study of pathogenesis is the key to the understanding of any given virus disease. For if all details of pathogenesis are understood, then one can deduce the likely clinical features, understand the method of spread of the infection, and comprehend the essentials of the immune mechanism. Furthermore, intelligent methods of control by means of vaccines and other measures can be formulated.

Experimental study

The investigation of pathogenesis in man is difficult, because the sequential spread of virus takes place in the incubation period, before a physician is called. Some relevant information has been obtained from study of volunteers and close contacts of cases of disease.

One of the classic studies of pathogenesis in the laboratory was conducted by Fenner who investigated infection with ectromelia, a naturally occurring poxvirus infecting mice. The virus closely resembles vaccinia and variola, and there are obvious lessons applicable to human infection with poxviruses. After inoculation of ectromelia virus in the foot pad, virus first multiplied in the regional lymphatic glands, being detected there as early as 8 hours after inoculation. Multiplication in the lymph nodes continued for a few days, and this was followed by a primary invasion of the blood stream, or viremia. Later, the virus proliferated in liver and spleen, and then supervened a period of secondary viremia. Virus probably escaped into the blood by direct release from necrotic cells. Finally, a focal infection of the skin occurred, and a generalized exanthem was noted on the ninth day after inoculation. Virus multiplied in the skin. The infection was terminated by death, or by the development of antibody at about the tenth day, which limited the spread of virus to new cells. In this study, it was found that the process of virus dissemination was sequential, and that stepwise increase occurred during the incubation period.

Returning now to disease of man, viruses may enter the body by one or more of several portals of entry, such as the upper respiratory tract, the conjunctival sac, the mouth, through the skin, or mucous membranes, or by direct implantation of an arborvirus into the blood stream by a biting arthropod. The nasopharynx is probably the commonest portal of entry of viruses. While in some instances, for example conjunctival infection and warts, subsequent proliferation of virus may be purely local, it is clear that in almost all infections spread to regional lymph glands takes place. This is probably sufficient to stimulate antibody production. It is not clear how frequently viremia occurs, but increasingly it is being found that blood spread is a common occurrence in virus infections.

It may be of interest to refer to the pathogenesis of a few selected virus infections. Thus, the pathogenesis of the arthropod borne virus infections is relatively simple. Presumably, the arthropod vector deposits the virus directly in the blood stream. Proliferation takes place in the internal organs before the stage of viremia. Eventually, virus localizes in target organs, especially liver, central nervous system, and skin.

Of recent years, the pathogenesis of poliomyelitis has been much studied, and many of the details have been clarified. Virus probably enters by and proliferates in the lymphoid tissue of the throat and Peyer's patches of the intestine. It then passes to the regional lymph nodes where further multiplication takes place. Virus then enters the blood and circulates for a few days, during the incubation period. This leads to the stimulation of blood antibody, which soon neutralizes the viremia. At about this time, virus may enter the central nervous system and spread in the brain and cord, probably by axonal routes.

It is almost certain that many factors can influence the spread of virus through the body and localization in target organs. The role of pregnancy, hormonal factors, and physical exhaustion is not too well understood, but some diseases in pregnancy tend to be severe, and muscular exercise influences the localization and course of paralytic poliomyelitis. There seems little doubt that corticosteroids tend to aggravate virus infections.

Tropism of Viruses

One of the interesting biologic properties of a virus is its tropism, that is to say its selective affinity for certain organs.

Tropism does not seem to be too closely related to other properties of a virus. Doubtless, tropism is genetically determined, and in that sense we can speak of tropism as a "marker."

During the course of laboratory manipulation, viruses alter in these properties and lose capacity to grow in certain organs. An excellent example is poliovirus, which by continued passage and selection, in animals, eggs, and tissue cultures loses neurotropism while retaining enterotropism. The 17D strain of yellow fever has lost viscerotropism while retaining a degree of neurotropism.

Broadly speaking, viruses can be classified into dermotropic, neurotropic, pneumotropic, enterotropic, and hepatotropic groups. Increasingly, however, it is realized that many virus strains are polytropic or pantropic, attacking many different organs. Coxsackie virus type B5 is an excellent example of a virus with such properties, for it causes myocarditis, pericarditis, myositis, hepatitis, aseptic meningitis, and encephalitis.

Histologic Changes

It is characteristic that changes induced by viruses strat in individually infected cells and spread out from individual foci. The most characteristic histology in a virus infections is degenerative change in epithelial cell. This occurs, for example, in epithelial cells of the skin in infection with poxviruses and herpesviruses. The cells become ballooned and eventually break down to form vesicles. Degeneration also occurs in the liver in yellow fever, Rift Valley fever, and hepatitis. Also under this heading comes the destruction of ganglionic nerve cells that occurs in poliomyelitis and encephalitis. Degenerative lesions also occur in myocardial and striated muscle fibers in Coxsackie infection. A special form of degeneration is that found when rubella virus attacks the developing cells of the fetus. Degeneration is also seen in a highly specific form when influenza virus destroys the ciliated epithelium of the respiratory tract.

Some viruses stimulate cell to proliferate and form new (tumour) growth. Examples include molluscum and warts viruses in the human skin, Shop's fibroma and papilloma in the skin of rabbits, contagious pustular dermatitis of man and sheep, Rous sarcoma of chickens, and polyoma of rodents (Table 10.1).

Table 10.1 Virus Induced Tumours in Animals

Species of Animal	*virus*	*Discoverer*
Rabbit	Fibroma	Shope (1932)
	Papilloma	Shope (1933)
	oral papillomatosis	Parsons and Kidd (1936)
Squirrel	Fibroma	Kilham (1953)
Deer	Fibroma	Shop (1955)
Dog	Oral Papillomatosis	Monbreun and Good pasture (1932)
	Venereal sarcoma	Smith and Washburn (1898)
Horse	Papilloma	Cook and Olsen (1951)
Cattle	Warts	Creech (1936)
Frog	Kidney carcinoma	Lucké (1934)
Chicken	Rouse sarcoma	Rous Rous (1910)
	Avian leukemias	Ellerman and Bang (1908)
	Avian myeloblastosis	Beard et al. (1952)
	Avian erythroblastosis	Beard et al. (1952)
Mouse	Mammary carcinoma	Bittner (1936)
	Leukemia (spontaneous)	Slye (1931); Furth (1935)
	Leukemia (induced by radiation)	Krebs et al. (1930)
	Friend virus	Friend (1956)
	Lymphoid-leukemia	Maloney (1959)
	Polyoma	Groos (1953(; Eddy et al. (1958)
Hamster	Polyoma	Eddy et al. (1958)
Rat	Polyoma	Eddy et al. (1959)

It has been said that hyperplasia induced by viruses is of three types. In the first, the proliferation is self limited, as in fowlpox and Shope's fibroma. In the second type, the virus causes a neoplasm, but disappears when malignancy results, as in Shope's papilloma of rabbits. In the third category is the Rous sarcoma of chickens, where the virus induces the tumour and persists in the tissues.

Inflammation

Although degeneration or proliferation are the specific histologic changes induced by viruses, some degree of secondary inflammatory reaction is almost always present. An outpouring of inflammatory cells surrounds the focus of viral action, and there may also be a more diffuse reaction. The inflammatory cells tend to be mononuclear rather than polynuclear. The formation of pus is unusual, except in lymphogranuloma venereum, unless there is secondary bacterial contamination. Scar tissue develops if there has been involvement of the dermis, as in variola, vaccinia, and varicella, or in the conjunctiva in trachoma. In the peripheral blood, lymphocytosis is a usual finding.

Myocarditis is now recognized to be fairly common in virus infections. It occurs in a severe form in Coxsackie B infection, but also in poliomyelitis, influenza, psittacosis, and generalized vaccinia.

Latent Infections

A special example of latent infection is lymphocytic choriomeningitis in mice. The virus is transmitted in *utero* to the offspring, and the baby grows up to become a carrier. Antibody does not develop, and this is an example of immune tolerance. Another prevalent virus of mice, Theiler's mouse encephalomyelitis, is spread to young mice early in postnatal life. They become latent carriers, and only a few develop paralytic illness.

Herpes simplex represents the best example of prolonged latency of a virus infection in man. Infection with this virus is common in early life, when vesicular lesions occur on the skin or in the mouth. Following clinical recovery, the virus survives in the body, perhaps in mucous membranes or lymph nodes, but only occasionally becomes awakened into activity, with the development of clinical herpes. This may be precipitated by endocrine, nutritional, or traumatic factors, or by feverish illnesses. Persons latently infected with herpes simplex virus show the presence of circulating virus neutralizing antibody, yet this does not eliminate the virus, suggesting that the virus may spread from cell to cell without being exposed to body fluids containing antibody. It is now realized that number of other viruses and rickettsia may persist in the body for prolonged periods following apparent recovery, e.g., varicella-zoster virus, psittacosis, probably adenoviruses, and *Rickettsia prowazekii* of epidemic typhus.

Virus Infections and Congenital Malformations

It was been known for many years that if certain virus infections occur in early pregnancy, abortion is likely to result (e. g., smallpox, poliomyelitis). A new chapter in the pathogenesis and pathology of virus disease was opened by the Australian ophthalmic surgeion Gregg when in 1941 he reported an "epidemic" of congenital cataracts in newborn babies, Inquiry disclosed that most of the mothers had suffered from rubella in the first 3 month of pregnancy. At this time the disease was epidemic in Australia, and there were many cases in young adults. Swan in South Australia, Ingalls in the United States, Hill and Manson in the United Kingdom, and others, have confirmed and extended these observations.

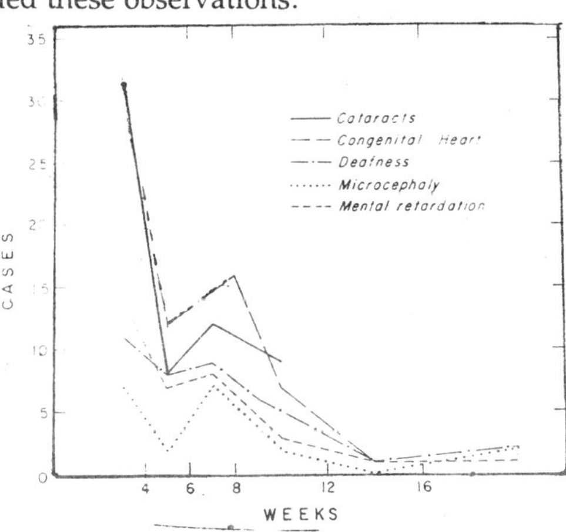

Fig. 10.1 Stages of gestation in relation to malformation. One hundred and eight patients who had various congenital abnormalities associated with maternal rubella during pregnancy are plotted in relation to the stage of gestation when damage occurred. [Reproduced from Dekaben et al., 1958. Neurology (Minneap.), 8: 387.]

It is now known that rubella in the first 3 months may produce the following defects; congenital cataracts; deafness due to damage to the organ of Corti; and congenital heart lesions such as patent ductus arteriosus and ventricular septal defect. Less common lesions include dental abnormalities, microcephaly, mental deficiency, and cleft palate.

These lesions are presumably caused by the virus of rubella damaging irreplaceable cells at the time of differentiation and early development. The virus may cross the placenta even in cases which appear clinically quite mild.

More detailed analysis of the sequels of maternal rubella has revealed that the virus is responsible for a whole series of departures from normal, namely, congenital malformations demonstrable at birth, or later in life (e.g., deafness and heart disease), abortion, stillbirth abnormally low birth weight, and death in the first year of life. The total wastage from rubella may be 50 per cent if the maternal attack is in the first month. The rate falls for rubella infection in the second and third months.

The teratogenic action of the still undiscovered rubella virus has been proven by epidemiologic inquiry beyond doubt. These observations have stimulated many workers to search for a similar action of other viruses. It may be said that no other virus infection in early pregnancy is followed by the triad of congenital cataract, inner ear deafness, and heart defects.

There is some evidence that attacks of influenza in early pregnancy may increase the incidence of anencephaly. Congenital defects of various types have been attributed to maternal poliomyelitis, Coxsackie infection, measles, and mumps, but the evidence is incomplete.

The widespread oral administration of live poliovirus Vaccine to women in the first trimester of pregnancy, should bring to light any teratogenic action of the virus, if it occurs. So far no association has been reported.

Neonatal Virus Infections

Viruses infecting the mother in the last few weeks of pregnancy may also infect the baby. In some instances, the virus crosses the placenta, and the infection is truly intrauterine. The baby is born with evidence of the disease in question, or develops signs within a day or two. Infections falling in this category include smallpox, measles, varicella, zoster, herpes simplex, poliomyelitis, probably hepatitis, and Coxsackie infection. Infection may also be transmitted to the baby at the time of delivery or very shortly thereafter. The infant develops signs of the disease in question after a somewhat shortened incubation period.

Of recent years, the greatest of interest has been taken in neonatal herpes simplex and Coxsackie group B infections. Both infections are characterized by a high fatality rate and generalized dissemination of virus in the body. Myocarditis is particularly characteristic of Coxsackie B infection.

Oncolysis Induced by Viruses

Several viruses have been found to grow in and destroy tumour tissue, e.g., Russian spring summer encephalitis, West Nile, Japanese B encephalitis, St. Louis encephalitis, Semliki Forest, louping ill. The oncolytic action of these viruses has been demonstrated in malignant chicken tumours, rabbit fibromas, and mouse sarcomas.

Oncogenesis

A large number of oncogenic (tumour-producing) viruses has now been isolated from animals, since the early discovery of avian lymphoblastosis (1908) and the Rous sarcoma (1910). It will be seen that the tumours involve rabbits, squirrels, deer, dogs, horses, cattle, frogs, chickens, mice, hamsters and rats. So far, warts and molluscum contagiosum are the only viruses recognized to produce tumours in man, and these are benign.

11

Immunity in Virus Infections

Natural Immunity

Several non-specific mechanisms operate to minimize the chances of viruses entering and multiplying in the human body. Much work, for example, has been done on the large category of non-specific viral inhibitors found in normal serum. In table 11.1 we reproduce the properties of the better known inhibitors (Allen and associates). It will be seen that these can be divided into heat stable and heat labile. The heat stable normal inhibitors are either protein, mucoprotein (Francis inhibitor), or lipid. The protein inhibitors neutralize infectivity and have been found in the serum of cattle, horses, monkeys and other animals. It is now realized that viruses, especially enteroviruses, are very prevalent in the animal kingdom, and it may be that the so-called non-specific neutralizing inhibitors are in fact antibodies developed in response to infection with viruses that have caused a latent infection.

The mucoprotein heat stable inhibitor or Francis (α) inhibitor act on hemagglutinating myxoviruses. It does not neutralize the infectivity of influenza virus, and in fact is destroyed by live virus, as well as by the receptor destroying enzyme (RDE), and periodate. This inhibitor prevents the hemagglutination caused by myxoviruses, and inhibits heated virus to a higher titer than live virus. It is similar in composition to the mucopolysaccharide red cell receptor substance.

Table—11.1 Spectrum of Activity of Viral Inhibitors Found in Normal Animal Sera

Type of Inhibitor	*Chemical Nature*	*Viruses Affected*
Heat stable (56°C. /30 min.)	Protein or probably protein	Adenoviruses Enteroviruses Poliovirus Echo Arborviruses Yellow fever Japanese E[a] St. Louis E West Nile
	Mucoprotein[b] Francis (α) inhibitor Others with nature unknown	Myxoviruses Influenza Mumps New castle disease
	Lipid	Myxoviruses Influenza Newcastle disease Arborviruses[c] Psittacosis
Heat labile (56°C. / 30 min.		
1. Neutralizing enhancing	Proteins and protein complexes	Myxoviruses Mumps Newcastle disease Aborviruses Western equine E Dengue Simbu Wyeomyia Vaccinia Rous sarcoma Lymphocytic choriomeningitis Measles Variola Herpes
2. Inactivating	Proteins and protein complexes (Chub, β inhibitors, properdin)	Myxoviruses Influenza Mumps Newcastle disease Arborviruses St. Louis E Murrary Valley E Wyeomyia Vaccinia Variola Herpes Bacteriophage (*Escherichia coli*) (Staphylococcus)

[a] E, encephalitis.

[b] Inhibitor affecting hemagglutinin primarily.

[c] Including Japanese, St. Louis, Russian spring summer encephalitis, and others.

There is also a group of heat labile inhibitors. For example, the neutralizing capacity of heated specific antiserum for certain viruses can be enhanced by the addition of complement. There are also heat labile inhibitors that neutralize virus *per se*. One of the best known is the Chu or β inhibitor which inhibits the hemagglutinating action of live and heated influenza virus as well as infectivity. This inhibitor is not destroyed by purified RDE or periodate. There is increasing evidence that heat labile non-specific antibodylike susbtances are in fact associated with the properdin system.

Many other factors play a role in natural immunity, but the significance of these is poorly understood. For example, suckling mice are susceptible to several viruses, but this susceptibility to infection is gradually lost with increasing age, the so-called maturation resistance effect. Most Coxsackie viruses infect only suckling and not weaned mice. Various arborviruses with a neurotropic property infect baby mice when injected by peripheral routes, probably spreading by nerve pathways. This susceptibility of the mouse is lost as age advances, although the inoculation of the same virus by the cerebral route still secures infection.

Hormonal factors probably play some role in determining resistance, and the evidence for this is the administration of steroids increases the susceptibility of animals to viruses such as enteroviruses and herpesviruses.

Active Immunity Following Infection

Resistance to reinfection with the homologous type of virus develops to some degree on recovery from a clinically evident attack of any virus disease. Broadly speaking, virus diseases tend to fall into two categories in regard to the duration of the resistance that follows infection. In the first category, virus disseminates widely through the body, there is viremia, and virus multiplies freely in the various affected organs. In these diseases, the incubation period is often long, over 10 days, so that there is opportunity for virus to spread through the body, multiply and give a powerful stimulus to the antibody-forming mechanism. Examples of diseases in which there is a prolonged resistance to reinfection include smallpox, vaccinia, measles, rubella, varicella, yellow fever, equine encephalitis, and mumps.

Diseases in which the duration of immunity is only transient include the common cold, influenza, sandfly, and dengue fevers. In some of these infections, the multiplication of virus occurs chiefly in the superficial mucous membranes, and the incubation period is short. These factors do not favour widespread virus multiplication. The stimulus to the antibody-forming mechanism is therefore reduced. It is possible that in some of these infections the true explanation of a repeat attack is invasion by a different antigenic type of virus, and the duration of immunity to the strictly homologous strain may be longer than generally believed.

There is good evidence that active immunity also develops on recovery from an infection that does not give rise to clinical disease, so-called inapparent or symptomless infection. This type of infection may be associated with widespread invasion of the tissues with virus, and with the formation of antibodies to high titer. Resistance may be as firm and lasting as that following clinically evident disease. Resistance is known to follow subclinical infections with mumps, yellow fever, many of the arbor viruses, hepatitis, poliomyelitis, and adenoviruses.

The prolonged, often lifelong, resistance that follows an attack of a virus disease is a very striking phenomenon, the precise explanation of which is still lacking. Almost certainly, the resistance depends on one or more factors, perhaps not all of which have yet been recognized. In some infections, the superficial tissues develop a degree of resistance to reinvasion. For example, the Superficial epithelium of the respiratory tract becomes stratified instead of cuboidal during recovery from influenza. Stratified epithelium is less susceptible to invasion than the usual type. Particular study has been made of the local resistance that develops in the gastrointestinal tract on recovery from infection with poliovirus and other enteroviruses. Local resistance limits the multiplication of virus on reexposure to infection. The reason for this limitation of multiplication, which is relative and not absolute, is not clear. It may well be that local development of neutralizing antibody in the lymphatic tissues of the pharynx and intestinal tract is a partial explanation.

Antibody formation is clearly the major mechanism that hinders the spread of virus through the body on re-exposure to infection. Antibody, and neutralizing antibody is obviously the one concerned, develops within 7 to 14 days of onset of most virus infections, and brings to an end any viremia that may have been present. Antibody may exert some effect in preventing multiplication in the mucous membranes and gastrointestinal tract as well as in lymph nodes and internal organs. It seems likely, however, that the major role of antibody is to prevent carriage of virus through the blood stream.

Neutralizing antibody may persist at a measurable level in the blood stream for many years. In other infections, the level falls, but a reinvasion often serves as a booster, and the antibody titer rises promptly to prevent further virus spread through the body. This prompt outpouring of antibody following re-exposure to virus has been held to constitute part of the immune mechanism in poliomyelitis (Salk).

The phenomenon of persistence of antibody has stimulated must discussion. In many instances, the continuation of a high antibody level may merely be due to repeated reinfecting doses of virus acting as boosters. This may be the explanation for continued high levels of antibody to poliovirus, measles, and other viruses which are widely prevalent. However, there are cases on record where high levels of yellow fever antibody have persisted long after the person has left the yellow fever zone and in the absence therefore of an opportunity for re-exposure. It is also possible that exposure to viruses sharing a common antigen may serve to maintain high antibody levels. In deed there is evidence that an explanation of this sort may be valid in the case of the influenza type. A viruses and arborviruses.

A major subject for discussion is the possibility that the continuation of the high antibody levels long after convalescence may be due to persistence of the virus concerned in the body tissues. That is to say, we may be dealing with an infection immunity. In the case of epidemic typhus, of course, we have the example of Brill's disease, in which the patient suffers from a relapse of typhus many years, perhaps a lifetime, after the original attack. The rickettsiae have been isolated from the lymph glands decades after the primary attack. Psittacosis and adenoviruses also seem to persist for some time after the acute attack. Infection immunity may, therefore, provide an explanation for the persistence of immunity.

Of course, one can speculate about the nature of viruses and point out that some form of antigen may persist indefinitely after the acute attack. The previous sharp dividing lines between dead virus and live virus have become less real. Poxviruses that have apparently been inactivated by heat can be reactivated in the presence of live virus. It is possible, therefore, that following recovery from a virus infection, effective antigen may persist indefinitely.

One may perhaps conclude this section by the reminder that although humoral factors seem to play the major share in the immune response, the role of tissue factors must not be overlooked. For example, children with congenital agammaglobulinemia, although showing little resistance to bacterial infection seem to have a normal resistance to viral infections; in these children, circulating antibody does not develop. Work on interferon suggests that the liberation of this substance from normal cells, under the action of virus, plays an important role as the first barrier against cellular invasion by virus.

Active Immunity Following Vaccination

Vaccines used in human and animal virology fall into two groups, the live virus vaccines and those vaccines containing killed or inactivated virus. There is general agreement that live vaccines are to be preferred to killed vaccines as they give a longer lasting immunity. Live virus vaccines produce their immunizing effect by proliferating in the body, generalizing, often producing viremia, and thus giving a widespread stimulus to the antibody-forming mechanisms. In effect, vaccine leads to a subclinical infection.

In man, because of some abnormal susceptibility of the host, the live viruses used in the prevention of smallpox and yellow fever may sometimes produce a frank clinical infection. Smallpox vaccination, for example, may be followed by a generalized vaccinial eruption. In veterinary medicine, this slight risk in the use of live viruses does not constitute a serious deterrent to employment. It is not a matter of importance if a small percentage of an inoculated herd or flock becomes sick or even dies, provided that the great majority becomes protected.

Live viruses for use in veterinary vaccines come from many sources. Some strains of Newcastle disease, fowlpox, and laryngotracheitis viruses recovered in the field are possessed of such low virulence for the host that they can be employed for immunization purpose. A commoner type of live virus is one derived from serial passage in different animal hosts, chick embryos, or tissue cultures. During this process mutant forms have an opportunity to develop, and gradually raplace the dominant parent strain. There are many examples of such attenuated viruses in veterinary medicine. The best known include African horse sickness, blue tongue, canine distemper, hog cholera, rabies (Flury), and rinderpest.

In the human field, the classic examples of the development of live avirulent viruses by repeated serial passage are vaccinia and yellow fever. Vaccinia has been derived from smallpox by repeated transfers in calves from rabbits. There are two strains of yellow fever used for vaccination. The 17D strain was developed by serial passage in tissue cultures, and the neurotropic strain by passage in mouse brains. These yellow fever virus strains have lost their capacity to invade the liver and cause hepatitis, but they do cause viremia. There is currently much interest in this field, with active studies in progress on attenuated vaccines for use in poliomyelitis and measles. Russian workers also have developed attenuated vaccines for use in influenza. The Flury rabies virus is now being used in man.

Attenuated poliovirus vaccines have been developed by serial transfer of virulent strains through rodents, eggs, and especially tissue cultures. Mutants with a low degree of virulence for the nervous system have been selected by these manipulations.

Avirulent viruses developed by the above technics of serial passage, or found in nature, appear to remain constant in their biologic properties and seldom revert.

As regards the human field, of the various vaccines now in fairly general use (Table 11.2) most fall into the category of killed vaccines. These are prepared from various sources: chick embryo fluids or tissues, as in the case of influenza, mumps, or equine encephalitis; brain and cord tissue, as in rabies; poliomyelitis and adenovirus vaccines are prepared in monkey tissue cultures. It is certain that in the not far distant future more vaccines prepared in tissue cultures will be introduced, especially for the widely prevalent respiratory viruses.

Virus suspensions obtained by these methods are treated by various chemical or physical means to destroy the virus. Ultraviolet light has been used to prepare rabies vaccines. Formalin is added to poliomyelitis, influenza, mumps, adenovirus, and equine encephalitis suspensions. In poliomyelitis much study has been devoted by Salk and others to the methods of destroying virus with retention of antigenicity. By adjusting the concentration of formalin and the time of exposure, a preparation can be obtained which does not contain detectable residual live virus, and which proves antigenic in experimental animals and man stimulating antibody and resistance.

Both live and killed vaccines immunize by the same means, that is to say, by the production of serum antibody. Live poliovirus vaccines also stimulate local resistance of the gastrointestinal tract. Antibody develops within about a week to 10 days of exposure to a live virus, increases to a peak, and then slowly fall. In yellow fever, antibody can be detected in serum many years following inoculation.

Following an injection of a killed virus vaccine, serum antibody also develops fairly quickly, but not so rapidly as following actual invasion of the body by a live virus. Following an injection of poliomyelitis vaccine, antibody reaches a titer of between 1:50 and 1:100 within a month after the injection. This level falls during the nest few months. If another injection is given some months following the primary course, antibody level rapidly rises. This increase occurs within a space of about 7 days, and is an example of the secondary response, or the so-called booster effect.

Vaccines are usually given subcutaneously or intradermally, but skin scarification is used for smallpox vaccination, and the Dakar type of yellow fever vaccine. In many cases, the effect of the initial does or doses of vaccine should be enhanced by a booster dose given several weeks later. Combined vaccines, especially those containing polio, diphtheria, pertussis, and tetanus prophylactics are being widely used.

It is well to remember that, in rare cases, vaccines containing animal and egg tissues may give rise to sensitization, with serious consequences; brain tissue vaccine may lead to demyelination.

The principal vaccines used in the prevention of disease in man are shown in table 11.2, and those used in veterinary practice in table 11.3.

Allergy

Allergic sensitivity develops during convalescence from a number of virus infections. This state can usually be demonstrated by the intracutaneous injection of heat killed virus preparations. A positive reaction is manifested by the formation of an erythematous macule or papule that may persist for a few days. The Frei reaction in lymphogranuloma venereum furnishes a good example of this type of response. Heated or otherwise inactivated virus is injected intradermally, and in infected persons a papule develops. Skin reaction can also be demonstrated in infection with the viruses of vaccinia, herpes simplex, and mumps. In general, a positive skin test is correlated with the occurrence of serum antibodies, and denotes previous infection. In mumps, the reaction to the skin test has been used a means of detecting children who are susceptible to natural infection. A good illustration of allergy is that found in vaccination against smallpox. When revaccination is carried out in persons possessing a high degree of specific resistance to vaccinia virus, an immediate reaction, characterized by the rapid

Table—11.2 Principal Vaccines Used in Prevention of Human Virus Infections

Diseases	*Source of Vaccine*	*Condition of Virus*	*Route of Inoculation*
Adenovirus infections	Monkey kidney tissue culture (formolized)	Inactive	Subcutaneous
Equine encephalitis (North American)	Chick embryo (formolized), containing one or both antigenic types	Inactive	Subcutaneous
Influenza	Chick embryo (formolized) containing various strains	Inactive	Subcutaneous
	Chick embryo and human transfer	Attenuated	Atomization
Japanese (B) encephalitis	Chick embryo (formolized)	Inactive	Subcutaneous
Measles	Chick embryo tissue culture	Attenuated	Subcutaneous
Mumps	Chick embryo (formolized)	Infective	Subcutaneous
	Chick embryo	Attenuated	Subcutaneous
Poliomyelitis	Monkey kidney tissue culture (formolized)[a]	Inactive	Subcutaneous
	Monkey kidney tissue culture[a]	Attenuated	Oral

(Contd...)

Table 11.2 (Contd...)

Diseases	*Source of Vaccine*	*Condition of Virus*	*Route of Inoculation*
Rabies	Rabbit cord (desiccated)	Attenuated	Subcutaneous
	Brains of animals exposed to formalin, phenol, chloroform, ether, or ultraviolet light	Active or inactive, depending on duration of exposure	Subcutaneous
	Flury strain in chick embryo	Attenuated	Subcutaneous or intradermal
Russian spring summer encephalitis	Mouse brain (formolized)	Inactive	Subcutaneous
Smallpox intradermal, (desiccated or glycerolated)	Lymph from calf or sheep	Attenuated	Scarification, or multiple pressure
Venezuelan equine encephalitis	Chick embryo (formolized)	Inactive	Subcutaneous
Yellow fever	Tissue culture and chick	Attenuated embryo (17D)	Subcutaneous or scarification
	Neurotropic virus (mousebrain); may be combined with vaccinia	Attenuated	Scarification

[a] The three types may be combined.

development of an erythematous papule, may be noted. This reaction may also be elicited by heated virus, and indicates sensitivity to the virus protein.

In general, it can be asserted that allergy does not play any significant role in the clinical manifestations of virus diseases. However, it is possible that second attacks of influenza, occurring a short while after the first attack, may be explicable on an allergic basis. There is mounting evidence that allergy may operate in some of the post-infectious and post-vaccination encephalomyelitides.

Serum Antibodies

Serum antibodies develop on recovery from virus infections, whether the illness in apparent or inapparent, and after the infection of vaccines. Antibody can also be found in the secretions of the respiratory tract.

It is beyond our scope to describe the theories relating to antibody formation. Many studies, including the use of fluorescent antibody (Coons), point to the liberation of antibody gamma globulin by plasma cells descended from lymphocytes that react with specific viral antigens (Burnet).

Antibody can be detected by several technics, such as agglutination, precipitation, complement fixation, neutralization of infectivity, and inhibition of haemagglutination. It is not known whether these technics measure different properties of one antibody, or different antibodies, but more likely the latter. In the period of active disease, there is usually fair correlation between the results obtained by different methods for the estimation of antibody. In specimens taken in the early days of infection, however, antibody may be detected by certain technics only, and in later stages, likewise, only some tests may prove positive. In general virus neutralizing antibodies persist in the serum for longer time than complement fixing antibodies.

The rate of development of antibody varies in different infections. Serum removed within the first few days of onset (the "acute phase" sample) usually shows no antibody or only a low level, but the level is higher in a second sample withdrawn about 10 to 14 days after the onset of illness (the "convalescent phase"

Table 11.3 Principal Vaccines Used in Prevention of Virus Infections of Animals

Diseases	*Source of Vaccines*	*Condition of Virus*
African horse sickness	Attenuated mouse passed virus	Attenuated
Canine hepatitis	Tissue vaccines	Inactive
	Porcine tissue culture, attenuated virus	Attenuated
	Ferret tissue culture, attenuated virus	Attenuated
Canine distemper	Chick tissue culture, attenuated virus	Attenuated
	Avianized (egg adapted)	Attenuated
	Ferret passaged virus	Attenuated
Contagious ecthyma	Dried scabs from infected sheep	Attenuated
Distemper of mink	Avianized (egg adapted) (1 does)	Attenuated
Enteritis of mink	Tissue suspensions (formolized)	Inactive
Equine encephalitis (North American)	Chick embryo tissue (formolized), bivalent or monovalent	Inactive
Feline panleukopenia ("distemper" of cats)	Tissue suspensions (formolized)	Inactive
Foot and mouth	Formolized tissue culture vaccine, produced in bovine tongue cultures	Inactive
Fowlpox	Chick embryo-propagated virus	Attenuated
	Chick embryo-propagated pigeon poxvirus	Attenuated

(Contd...)

Table 11.3 (Contd...)

Diseases	*Source of Vaccines*	*Condition of Virus*
Hog cholera	Tissue cultured virus injected simultaneously with hog cholera serum	Attenuated
	Crystal violet: defibrinated virulent blood from sick hogs, inactivated by crystal violet	Inactive
	Tissue vaccine: spleen, lymph nodes from sick animals, treated by addition of 1% eucalyptol	Inactive
	Lapinized: virus attenuated by serial passage through rabbits	Attenuated
Infectious laryngotracheitis	Chick embryo, chorio-allantois	Attenuated
Louping ill	Sheep brain and cord (formolized)	Inactive
	Allantoic fluid and chick embryo (formolized)	Inactive
Newcastle disease of poultry	Chick embryo (formolized) plus adjuvants	Inactive
	Avirulent strain produced in chick embryo for vaccination of day old chicks (intranasal)	Attenuated
	Avirulent strain produced in chick embryos (wing web inoculation, for older birds	Attenuated
Rabies	Sheep, goat, or calf brain treated with chemicals or ultraviolet light	Attenuated or inactive[a]
	Avianized (Flury) virus	Attenuated
Rinderpest	Dried goat spleen	Attenuated
	Spleen: virus inactivated with chloroform	Inactive
	Avianized	Attenuated

[a] According to duration of treatment.

sample). The rise in titer of antibody with the progress of convalescence is of the greatest value in establishing a laboratory diagnosis, but in some infections (e.g., lymphocytic choriomeningitis) the production of antibody may be slow and delayed for several weeks.

Antibody usually remains at a high level in the serum for a few weeks, and then slowly falls. In many cases, it may persist in measurable quantity for a prolonged period, even life, as after measles or yellow fever.

Production of Antiserum for Serologic Tests

In serologic work in virus infections, it is necessary to have available a stock of positive control immune sera known to contain specific antibodies, so as to provide a standard of reference for the testing of unknown viruses. These sera may be obtained from human beings convalescent from the disease in question, or animals recovered from experimentally induced disease. More usually, however, the sera are produced by immunization of animals with preparations of live virus. Antisera are produced in rabbits, mice, hamsters, ferrets, monkeys, chickens, and other animals. The rate of antibody formation varies in different infections. For instance, two injections of virus-infected allantoic fluid in the peritoneal cavity of a rabbit will suffice to stimulate a high level of neutralizing antibody to influenza, whereas to produce antiserum to vaccinia in the same animal, repeated intravenous injections may be required. Antibody response to certain viruses, notably rabies, influenza, and poliomyelitis, is greater if the antigen is incorporated in paraffin oil adjuvant (Freund). This is probably dependent on an increase in antibody liberated by cells in many parts of the body rather than on a local increase in antibody formation.

Virus antibody is mainly concentrated in the gamma globulin fraction of plasma.

Agglutination

This technic is relatively little used in virus work, but antisera will agglutinate the washed elementary bodies of vaccinia, fowlpox, psittacosis, herpes zoster, and varicella. The tests may be set up on slides, and read microscopically by dark-field illumination. Alternatively, narrow agglutination tubes can be used, small portions of fluid being withdrawn for microscopic, examination. With vaccinia virus, a sufficiently dense preparation of EBs can be obtained from the scarified skin of the rabbit's back for the clumping to be visible macroscopically. The agglutination of psittacosis EBs prepared from allantoic fluid is visible to the naked eye.

Precipitation (Flocculation)

It has been known for many years that when a suspension of poxviruses and antiserum are mixed in narrow agglutination tubes, flocculation takes place. This is in part a true precipitation between the soluble antigens, LS and others, of the virus and the antiserum, and in part an agglutination of elementary bodies. This type of reaction occurs with the following viruses and appropriate antiserum: variola, vaccinia, and other poxviruses, herpesviruses, influenza, mumps, Newcastle disease virus, and poliomyelitis. However, the test is little used except in the poxvirus group. The precipitin test is now almost always done by the gel diffusion (boundary) technic. In the Oudin single diffusion method, antigen and antibody are allowed to diffuse into a gel column. Opaque bands of precipitate develop, corresponding to the soluble antigenic components in the virus suspension. This has proved to be an elegant method of detecting minor differences in antigens between closely related members of the Poxvirus group.

In the Ouchterlony double diffusion method the reagents are placed in suitably spaced cups in a layer of agar. The precipitin bands form as distinct curved lines, depending on the number of antigens.

Fluorescent Antibody

The studies of Coons and his associates have had far reaching effects on both academic and diagnostic virology. They showed that viral antibody can be labeled with isocyanate of fluorescein or the isothiocyanate, and used to detect the presence of virus in cells. When antibody and specific antigen react, there is a brilliant greenish yellow fluorescence visible when the preparations are examined microscopically under illumination provided by an arc lamp or mercury vapour lamp. This technic has been used to study the site of antibody formation, as well as in the investigation of the sites of virus multiplication. It is also used extensively in diagnostic virology as a rapid means of identifying virus antigen.

Complement Fixation

It is necessary that a virus antigen used in complement fixation tests contain as much virus and as little host tissue as possible, in order to reduce non-specific fixation. The following types of antigen have all been used with success: suspensions of lung, spleen, muscle or brain, vesicle fluid, skin crusts, washed elementary bodies, tissue cultures, amniotic or allantoic fluid, ground chorio-allantoic membrane, yolk sac, or whole chick embryo. In the past, antigens have frequently been used in the live form, but it is safer to use non-infectious material, In psittacosis, the antigen is so heat stable that preparations may be boiled. Satisfactory non-infectious antigens have been prepared by extraction and precipitation with chemical calls, for example, alcohol, or acetone and ether, and by ultraviolet irradiation.

With several viruses, it has been demonstrated that there are two antigenic components that fix complement, the elementary body, virus or V antigen, and the soluble substance or soluble antigen of much smaller size. Soluble antigen are approximately 20 mμ in size and consist of nucleoprotein (RNA or DNA). They can be liberated from the elementary body and are not infectious. Soluble antigens are often group reactive and are shared by several members of the same virus group. Soluble antigens stimulate the production of antibody that reacts by the flocculation and complement fixation tests, but does not neutralize infectivity. Soluble antigen is more abundantly formed in some tissues than others, for example in the chorio-allantois and embryo lung.

Virus antigens are protein in nature and comprise the shell of the elementary body. They are presumably composed of the protein subunits (capsomeres). Virus antigens stimulate the development of antibody that react with intact elementary bodies by the agglutination, complement fixation, virus neutralization, and hemagglutination inhibition technics. In the case of influenza type A human virus, it has been estimated that there may be up to about 20 antigenically distinct virus antigens, one of which occurs in much larger amount and determines the subtype of the strain. Virus antigens may be shared between closely realted strains of a type, buy there is little evidence of any wide crossing.

In the conduct of the test, various methods may be followed, but a common practice is to prepare dilutions of serum, add a standard quantity of complement and a standard dose of antigen. Fixations is carried out at 37°C., or by placing mixtures overnight in the icebox (4°C.). After fixation, the hemolytic system is added, and the test incubated at 37°C. Alternatively, the dose of complement is varied instead of the serum, and here the degree of positivity of the test is expressed in terms of the number of units of complement fixed. Less commonly, the dose of antigen is varied, serum and complement being kept constant

Tests are now commonly performed by a technic introduced by Fulton in which the reagents are mixed in small hollows in a plate of transparent plastic material; incubation is carried out in a tightly sealed box to prevent evaporation.

In all work of this nature, it is essential to have adequat controls, because the antigens are prepared from tissues, and there is an ever present risk of the occurrence of nonspecifi fixation with normal tissue protein. It is necessary, therefore, to have an antigen control consisting of the same tissue uninfected by virus. The chances of non-specific fixations are reduced by inactivating all sera for 20 minutes at 60 or even 65°C., instead of the usual 56°C.

Complement fixation is used in the laboratory diagnosis of a great many virus infections. It is a considerable convenience to smaller laboratories that many suitable antigens are now available from commercial houses. In serologic diagnostic tests, it is desirable to examine in the same batch of fixation tests, acute and convalescent phase sera. According to the infection, serum titers of from 1:8 to 1:64 or more may be expected in convalescence.

Fixation tests are used in a somewhat different way in the diagnosis of smallpox, when antigen is prepared from the crusts or vesicle fluid of the suspected case, and tested for fixation with known antisera. In such tests, the appropriate controls consist of tissue crusts from a disease other than smallpox, e.g., varicella.

In addition to these applications, complement fixation is used in studying the antigenic structure of influenza and other viruses, and in identifying unknown viruses.

Of recent years, the indirect test for inhibiting antibody has been used in the serologic investigation of psittacosis.

Virus Neutralization

When preparations of virus are mixed with appropriate antisera, and the mixtures inoculated in susceptible hosts, infection, does not usually develop, because of the action of a virus neutralizing antibody. In these tests, it is the customary procedure to hold the mixtures of virus and antibody for 1 to 2 hours before inoculation, to allow time for neutralization, although with some viruses it is probable that neutralization may occur soon after the mixtures are prepared. The exact mechanism of neutralization of virus by immune serum is obscure, but is unlikely that the infectivity of all or even many of the virus particles acted on by the serum is actually destroyed. It is possible to recover active virus from such apparently inert virus serum mixtures by simple dilution, centrifugation, or adsorption of virus, suggesting that virus and serum are linked together in a loose combination.

The rate of neutralization of a virus varies directly with the concentration of antibody, and is most accurately measured when antibody is in great excess; the reaction then follows first order kinetics. Figure 11.1 shows the neutralization with time of type 1 poliovirus. The slope of the first portion of the curve is a measure of the rate of poliovirus neutralization, and is a measure of the probability per unit time that any given virus particle is neutralized by its antibody at a given concentration (Dulbecco). The decreasing rate of neutralization observed in the last part of the curve is due to the changing ratio of virus: antibody, and illustrates the existence of what has been called the "persistent" fraction of unneutralizable virus. Early in the modern history of virology, Andrews and Elford stated one of the major principles of virus neutralization. According to their theory, if antibody be in excess, a constant percentage of added virus remains active, regardless of the amount of added virus.

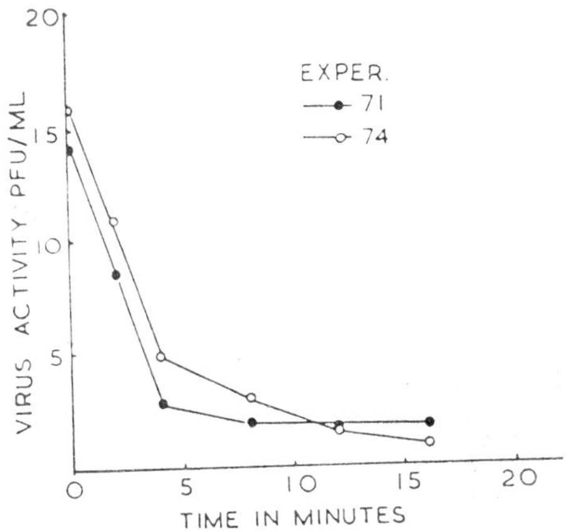

Fig. 11.1 Two experiments illustrating the neutralization of poliovirus type 1(Newfoundland strain) with homologous antiserum, dilution factor 10^{-4}.

The source of the persistent fraction was discussed many years ago by Burnet, who suggested that it was due to dissociation of active virus from the inactive virus-antibody complex. Later investigations have identified the persistent fraction with a proportion of the virus population that is resistant to neutralization. This resistance to neutralization is probably the result of phenotypic antigenic inhomogeneity of the virus population, and does not seem to reflect differences in virus genotype.

In a physiologic environment, with stable temperature, dissociation of virus from neutralizing antibody does not occur to a significant degree. Neither virus nor antibody is irreversibly changed by the union, for poliovirus may be reactivated by decreasing the pH of the medium after it has been combined with antibody for several months at 5°C. In other experiments, it has been shown that ultraviolet or heat inactivated virus can displace virus from the virus-antibody complex. This displaced virus is fully infectious and does not appear to be altered.

These experiments lead to the conclusion that virus neutralization is essentially a reversible reaction, the equilibrium constant of which greatly favours the formation of the virus-antibody complex.

Evidence has accumulated supporting the view that antibodies are heterogeneous in nature. This phenomenon affect virus neutralization in that the antibody produced early in response to a virus antigen is more specific than antibody produced later in the infection. This facet of virus neutralization must be considered if this reaction is to be used to differentiate strains of virus of the same type. This differentiation requires antisera of high specificity, but not necessarily of high titer.

The antigen used in these tests consists of fully active virus in the form of a tissue suspension, egg fluid, or tissue culture fluid. Preparations of elementary bodies can also be employed.

There is some evidence that virus neutralization induced by fresh serum is partly dependent on a nonspecific complement like component. This component retains activity when frozen, but activity is lost when storage is carried out at ordinary refrigerator temperature. This non-specific neutralizing effect can be minimized by the use of inactivated sera.

Titrations of antibody can be carried out by two methods. In the first, a fixed quantity of serum is employed, and falling dilutions of virus. The virus serum mixtures are held at room temperature for 60 minutes or longer, prior to inoculation in groups of 10 or more animals, cultures, or fertile eggs; it is not desirable to incubate the mixtures. Acute and convalescent sera are tested simultaneously, so that, in effect, the procedure is to titrate the virus in the presence of acute and convalescent phase sera. If the virus is one that kills the animals, the LD_{50} of the virus in each series is calculated. If death is not criterion of infection, the ID_{50} is employed. the LD_{50} or ID_{50} is expressed in logarithms, and the figures for each series of titrations are subtracted; the antilogarithm of the difference is known as the neutralization index. For example, if the LD_{50} of the virus in the acute phase sample is $10^{-6.5}$ and in the convalescent phase sample $10^{-4.0}$, the difference is $10^{-2.5}$ and the neutralization index therefore 320. The greater the difference in antibody level between the acute and convalescent phase sera, the higher is the neutralization index. Indices of over 50 usually indicate infection, and the figure may be considerably higher. Indices between 10 and 49 are inconclusive, and the tests should be repeated. Indices below 10 are regarded as negative.

The second method of titration involves the use of a fixed amount of virus, sufficient to kill 100 per cent of controls, with progressive dilutions of serum, for example, from zero by tenfold steps to 1:100,000. Both acute and convalescent phase sera are tested at the same time, and the mixtures inoculated in animals, cultures, or eggs. By this method, the 50 per cent protective or neutralizing endpoint of the serum is calculated, both for the acute phase and the convalescent phase sample. In a typical case of a virus infection, the 50 per cent protective point of the acute serum would be zero, and in convalescence the titer would rise to $10^{-2.0}$ or often much higher.

The mixtures of virus and serum can be tested for infectivity in many ways. With the neurotropic viruses, inoculations are usually made intracerebrally in mice. This, however, is a severe test, and may fail to detect small increase in antibody. A more sensitive method is to inoculate the mixtures by the subcutaneous or intramuscular route in baby animals. With other viruses, inoculations may be made in the skin, peritoneal cavity, testis, or other susceptible site. In tests of this type, it is necessary to keep the inoculated animals under observation for a long time, as serum may prolong the incubation period. The endpoint is taken as death, or the production of readily recognized lesions such as skin papulation, pneumonia, or paralysis, depending on the biologic characters of the virus in question.

It is now customary to substitute eggs for animals in work with a number of viruses. For example, fowlpox and laryngotracheitis antisera may be titrated by their capacity for inhibiting the developing of specific pocks on the chorio-allantois. Influenza and mumps antisera may be tested by inoculation into the amniotic or allantoic cavities.

The introduction of the modern types of tissue culture has enormously facilitated the conduct of tests for virus neutralizing antibody. Antibody titer is expressed in terms of the dilution inhibiting the cytopathic action of the virus in 50 per cent of the tubes inoculated. This technic has been widely used in the conduct of testes for poliomyelitis antibody in the serum of vaccinated persons and of suspected cases of the disease.

An ingenious development of the tissue culture technic introduced by the Maitlands and later developed by Enders has been perfected by Salk and his associates. This technic depends on the fact that uninfected kidney suspension metabolizes and produces acid. When virus is added, cell metabolism ceases, and the culture fluid remains unaltered. This test can therefore be employed not only for the titration of virus but also of antibody. Tests are conveniently set up in plastic disposable sheets.

Increasingly, use is made of the plaque reduction technic. Antibody reduces the number of plaques in monolayer cultures produced by a virus suspension in comparison with a control preparation from which antiserum has been omitted.

A special form of neutralization test is becoming popular with myxoviruses. Infected tissue cultures adsorb guineapig red cells and give the hemadsorption reaction. This hemadsorption is inhibited in the presence of specific antiserum.

Hemagglutination Inhibition

Certain viruses, for example, influenza, mumps, Newcastle disease, parainfluenza, arborviruses and vaccinia-variola, agglutinate various species of erythrocyte *in vitro*. This effect can be inhibited by the addition of immune serum. The test can be applied on a quantitative basis, by using a fixed amount of virus and red cell suspension and varying the dilutions of serum.

Passive Immunity and Serum Therapy

Under certain circumstances, passive immunity can be demonstrated in virus infections. One examples is found in the transient resistance of young infants to a number of diseases, associated with the transfer of maternal antibodies. Passive immunity can also be demonstrated following the administration of immune serum. For instance, if serum is given to an animal prior to the inoculation of virus, the normal progress of the infection may be aborted or prevented. If serum is administered shortly after the inoculation of virus, as in vaccinial infections, the same results may be noted. Even the clinical course of certain neurotropic infections can be influenced by the use of serum. For example, it was found that following inoculation of young mice on the foot pad with herpes simplex virus, serum could act on virus not only before it entered the epithelial cells, but even appeared to exert an inhibiting effect when invasion of the nerve fibers and central nervous system had occurred. In experimental rabies, it has been shown that serum administered shortly after infection may arrest spread of virus.

Turning now to disease in man, we find that serum has been used in a number of infections either as a prophylactic or curative agent. As a curative agent, serum is frankly disappointing. Serum has been claimed as effective in the treatment of postvaccinal encephalitis, a rare complication of antismallpox vaccinations.

As a prophylactic agent, antiserum finds its main application in the control of measles. Convalescent serum administered to an exposed person within a few days will usually prevent the development of an attack of measles; this state of passive immunity is of course transient. If serum is given toward the end of the incubation period, a modified attack of measles will often develop, and the immunity resulting therefore is permanent. As a source of serum, one may use the blood of convalescent adults, or pooled normal adult serum, because many persons have had measles in childhood.

Interest in the employment of antibody as a prophylactic has been greatly stimulated since the discovery that gamma globulin contains virus antibody concentrated severalfold over the level in serum. Gamma globulin prepared from the blood of normal blood donors contains contains antibody to measles, poliomyelitis, infectious hepatitis, herpes influenza, and other infections. The use of gamma globulin as a prophylactic is practically restricted to the first three of these, as discussed in the appropriate chapters, Gamma globulin prepared from the blood of convalescents has been recommended for administration to women exposed to rubella in the first trimester of pregnancy.

Hyperimmune gamma globulin has been prepared from the blood of volunteers inoculated with vaccinia. This product is being used increasingly in the prevention of smallpox, and in the treatment of vaccinial complications, especially in cases of dysgamma-globulinemia.

12

Industrial Microbiology

Learning objectives

- Be able to explain the three fundamental industrial processes.
- Describe the differences between batch and continuous culturing.
- Be familiar with the method used to culture *Saccharomyces cerevisiae* and the goals of such cultures.
- Be able to explain alcohol fermentation as it is carried out in the production of wine, beer, and liquor.
- Name several microbially produced enzyme products and describe their production.
- List the microbes utilized in the production of various vitamins and amino acids.
- Explain the microbial production of antibiotics and steroids.

Since prehistoric times, humans have taken advantage of the beneficial activities of microbes. However, it has only been within the past forty years that these activities have been harnessed for the large-scale production of microbial cells or their products. Microbiologist, engineers, and businesses have come together to develop the field of industrial microbiology. By controlling the activities of certain microbes, they have made possible the simpler, more economical production of large quantities of useful products such as enzymes, amino acids, vitamins, antibiotics, organic acids and alcohol.

Yeast, bacteria, and molds are grown in containers with capacities are large as 50,000 gallons. For example, yeast cells are grown for use in ironized yeast tablets for human use, and in even larger quantities as a supplement to farm-animal feed. It is not an easy job to handle large quantities of growing microbes. Expensive equipment and well-trained personnel are essential to the smooth operation of an industrial process such as antibiotic or beer production. Industrial microbiologists must closely monitor the chemical activities of the microbes and be prepared to change or stop the organism's activities to prevent product loss. Careful attention must also be paid to ensure that only one kind of microbe is being grown. Successful control of these factors has resulted in the development of multimillion-dollar microbial industries.

Cultivation of Microbes

Their small size, short generation time and rapid metabolic rate make microbes highly productive. Controlling microbial metabolic acclivities has enabled industrial microbiologists to develop three fundamental industrial processes. The first is the *cultivation* and *harvesting* of cells that may be used as food, food supplements, or inocula for other industrial processes. An example of this type of industrial process is the production of activated dry yeast and ironized yeast tablets. The second type of process utilizes select microbes as "factories" that enzymatically convert a particular substrate into a desired product, a process known as *bioconversion*. Steroids known as estrogens and progesterones used in birth control pills are produced by the action of microbes on specific substrates added to culture medium. The third industrial process utilizes microbes for their ability to generate large quantities of a specific *metabolic byproduct*. Alcohol, enzymes, and many types of organic acids are commonly produced by the microbial fermentation of inexpensive and readily available substrate material.

The industrial cultivation of microbes is carried out in large tanks known as fermenters. These units range in capacity from 5000 gallons (about 20,000 liters) to 50,000 gallons (about 200,000 liters). Some of the more basic and inexpensive substrates used as culture media are molasses (the thick syrup made from boiling sweet vegetables, sugar cane, or fruits), soybean meal, and malted grain. Fro specific bioconversion processes, more expensive synthetic media may have to the prepared from large quantities of purified ingredients. The fermenter is inoculated with a volume of culture equal to 5 to 10 per cent of the tank's volume. This means that a 15,000-gallon tank requires about 1500 gallons of pure culture to start the fermentation. This inoculum is built-up from a pure agar slant culture of microbes maintained in the laboratory, or it may be bought from a microbiological supply house. The initial culture is inoculated into ever increasing amounts of media to reach the final volume necessary for the processes. This repeated subculturing requires a great deal of space and close control to assure that contamination does not occur. In order to reduce these problems, lyophilized cultures can be purchased from supplies who have concentrated the required volume into a more manageable container. Once the substrate has been inoculated, all the conditions necessary for growth and reproduction must be accurately maintained. The optimum temperature, pH, purity, salinity, and cell density must be checked regularly. Since many industrial processes are aerobic, an adequate air supply must be made available to these fermenting microbes. This is done by injecting sterilized air into the tank through inlet valves at the bottom of the fermenter and continuously stirring the culture with mechanically operated paddles installed inside. The rapid bubbling and stirring causes a layer of foam to form on the surface that is controlled by the addition of antifoaming chemicals.

The two different methods that have been developed for industrial fermentations are *batch* and *continuous* culturing. Batch cultures are made by adding the microbes to a limited quantity of medium and allowing the fermentation to run to completion without the addition of fresh nutrients or removing by products. Cheese is made by batch culturing. Once inoculated into the fermenter, the microbes reproduce and increase in number according to the population growth curve. The maintenance of cultures in the exponential (log) phase of growth is known as steadystate, or balanced, growth and allows microbes to be cultured on a continuous basis. Continuous cultures are maintained in an apparatus called a *chemostat*. Many antibiotics, organic acids, and steroids are produced by microbes in continuous culture. The log phase of growth is maintained by continuously adding fresh, sterile medium to the *fermenter* at a rate equal to the removal of culture and products. The exponential growth rates regulated by maintaining the precise level of a limiting factor in the medium. This practice keeps the cells from shifting into a lag, or stationary, phase. Continuous cultures have an advantage over batch cultures, since the product can be constantly manufactured in large amounts. However, minor chemical or physical changes in the medium or spontaneous mutations within the microbial population prevent this method from being carried out longer than a few days.

1. Name the three fundamental types of industrial microbiological processes.
2. What are the main differences between batch and continuous culturing?
3. Microbes in a chemostat should be maintained in what growth phase?

Once the fermentation is complete, the product must be separated from the culture medium and purified. Passing the culture through a filter separates the medium from the microbes. If cells are the desired product, they are washed and concentrated by centrifugation. If the product is dissolved in the medium, it is separated by mixing with a solvent which selectively extracts the product from unwanted compounds. Depending on the products type, it is then separated from the solvent and crystallized in pure form. In some cases, the chemical may not be suitable in a crystalline form and must be redissolved for chemical modification. Members of the penicillin family of antibiotics result from chemical modifications of the basic penicillin molecule crystallized from such a process.

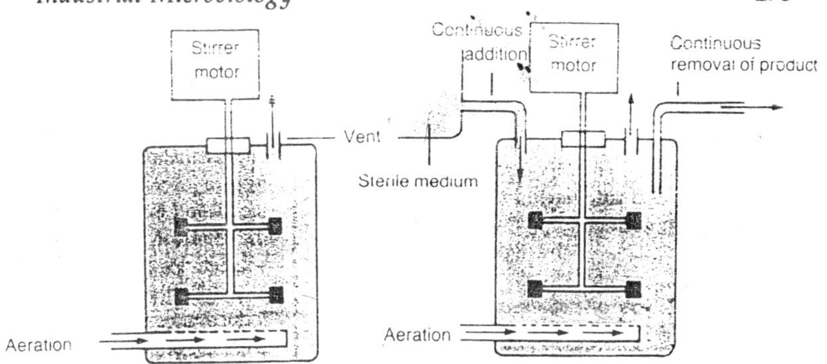

Fig.12.1 A typical batch fermenter (left) and a continuous culture unit (right).

With an increased understanding of organic chemistry, it has become possible to selectively modify many microbially produced compounds to the exact chemical structure needed for a particular purpose. However, successes in the chemical industry have not always benefited industrial microbiologists. Many of the organic compounds through to be produced only by microbes are now being manufactured in the chemistry lab. These products are manufactured in larger quantities from low-cost starting materials in less sophisticated equipment, enabling them to be marketed at a lower price. To overcome this problem, industrial microbiologists constantly search for new microbes or techniques to produce complex, commercially important compounds. One method used to increase production and reduce costs has been the development of productive mutations. *Productive mutants* are microbes that have been produced by natural selection following exposure to mutagenic levels of radiation. This technique was successfully used to increase penicillin production from the fungus *Penicillium chrysogenum* during World War II. Spores of this mold were isolated and irradiated with high dosages of ultraviolet light which caused mutations in the genes. When some of these spores were germinated, the new mycelia were found to produce larger amounts of the antibiotic than the original.

Developing an Industrial Process

In developing an industrial process based upon the action of micro–organisms many details must be given consideration.

Important among these are the type of culture needed, cultural conditions and the adaptability of the organism to large-scale production.

Purity and Nature of Cultures : It must be ascertained whether absolutely pure cultures must be used, or whether the mere predominance of one organism is sufficient. This may be a deciding factor, as the cost of preparing and maintaining pure cultures and sterile apparatus throughout a process is relatively high.

Cultural Conditions : The organism must be able to grow well in the medium to be used and under the conditions of the process. This necessities exact studies and careful control of *optimum* conditions of aerobiosis or anaerobiosis, temperature, nutrition and pH. Appropriate adjustments of the process and apparatus must be made to provide the necessary conditions.

Productive Mutants : The organisms selected must be such as will produce the desired substance (*s*) or results in the medium under the conditions furnished, in amounts sufficient to yield a profit. Some firms have "per" strains of micro-organisms that excel in producing certain products, such as butyl alcohol, certain antibiotics or itaconic acid, that they have "developed" (selected mutants) for these purposes. It has been found possible to induce industrially valuable mutations in micro-organisms by ultraviolet radiations. Where sexual processes are known to occur, the breeding of yeasts and molds for similar purposes is analogous to breeding of farm animals for special purposes.

Medium or Raw Material : The substrate or medium should support luxuriant growth of the organism to be used and it must be available constantly at costs compatible with profit. Expensive handling machinery may be needed for some substrates.

An important item is the possible necessity of a preliminary treatment such as liming of very acid yeast slops, distillery wastes, molasses and whey. Some substrates, such as sawdust or fiber, may need preliminary "digestion" with hot acid or alkali to hydrolyze them to fermentable substances. This all adds to the expense and time.

Nature of the Process : The more complicated and exacting the system of cultural details and preliminary heatings, dilutings and digestions, as well as the type of machinery (cracking stills, tanks, pumps) to handle the end-and by-products and the final wastes, the greater will be the cost and therefore the less the commercial practicability of any process. Any time-consuming aging or ripening process eats into profits. Sometimes, very desirable end or by-products may be found in commercial fermentations, yet the cost of their recovery may be prohibitive.

Preliminary Experimentation : The microbiologist working with 10-ml. test-tube cultures may find many valuable things. When attempts are made to reproduce the test-tube experiments on a 100,000 gallon factory scale, however, the laboratory discoveries often fail to yield the promised result. Any process developed in the experimental laboratory must next prove its worth in the factory. A small-sized model, or pilot plant, is usually tried after the preliminary laboratory work. All may depend on such a seemingly far-removed detail as international relations. These may affect the cost of importation of some raw product essential to the process under investigation. Then the industrialist turns to home resources, goes to Washington, or *employs a resourceful microbiologist!*

The whole matter is a complex of micro-biology, chemistry, engineering and economics. Only the micro-biology can be discussed here. Many chemical and micro-biological processes in use at present are patented and secret, and specific strains of bacteria, yeasts and molds, which are zealously guarded, are often carefully developed in the laboratories of manufacturing concerns. As a result of continuous and intensive industrial research, methods change or are superseded frequently.

Types of Fermentation Processes

Industrial fermentation processes may be divided into two main types: (1) batch fermentation and (2) continuous process. There are various combinations and modifications of these.

Batch Fermentations : A tank or fermentor is filled with the prepared *mash* (material to be fermented, e.g., diluted molasses, comminuted potatoes, digested corn cobs). The proper

Table—12.1 Some microorganisms important in industry

Micro-organisms	*Raw Materials (Substrates)*	*Commercially Valuable Products*	*Nature of the Process*
Bacteria:			
Clostridium aceto *bytylicum; Cl. saccharoacetoperbutylicum*	Carbohydrate mashes; molasses	Butanol, acetone, ethanol, vitamin B_2 (riboflavin)	Fermentation
Lactobacilius delbrueckii	Various wastes containing glucose or lactose	Lactic acid	Fermentation
Bacillus subtilis	Various organic wastes of dairies, canneries	Amylaces, protease	Aerobic or aerated tanks; synthesis
Bacillus polymyxa	Carbohydrate mashes	2-3-butanediol	Aerobic; metabolic waste
Leuconostoc mesenteroides	Waste carbohydrate mashes	Dextran	Fermentation
Leuconostoc citrovorum	Dairy wastes with citrate	Diacetyl	Fermentation

(Contd...)

Table 12.1 (Contd...)

Micro-organisms	Raw Materials (Substrates)	Commercially Valuable Products	Nature of the Process
Acetobacter sp.	Alcoholic liquors	Acetic acid (vinegar)	Aerobic; oxidation
Acetobacter suboxydans	Yeast extract with glucose, sorbitol, glycerol	Dihydroxyacetone, 5-ketogluconic acid, sorbose	Aerobic metabolic wastes
Streptomyces sp.	Corn-steep liquor	Antibiotics	Aerobic; synthesis
Yeasts and molds:			
Saccharomyces cerevisiae	Various mashes of grain, molasses	Industrial alcohol, alcoholic beverages	Fermentation
Aspergillus niger	Carbohydrate mashes	Citric acid	Aerobic dissim; aerated tanks
Penicillium notatum	Corn-steep liquor	Penicillin	"
Rhizopus negricans	Corn-steep liquor and progesterone	11a-hydroxypro gesterone	Aerobic; hydroxylation
Curvularia lunata	"	Corticosterone	Aerobic; dihydroxylation
Fusarium moniliform	Corn-steep liquor	Gibberellin	Aerobic; synthesis

Table—12.2 Some Industrial enzymes; their sources and applications

Type of Enzymes	*Source Micro-organisms*	*Typical Industrial Applications or Products*
Amylases (starch hydrolysis)	Malt *Aspergillus sp.*, *Bacillus* sp.	Bread; beer; whiskey; textile fibers; Preparation of glucose syrups
Proteases (protein hydrolysis)	*Aspergillus* sp. *Bacillus* sp.	Clarifying *(chillproofing)* beer; whiskey making, leather, baking bread and crackers, meat tenderizers
Pectinases (hydrolysis of pectin)	*Aspergillus* sp.	Clarification of fruit juices
Glucose oxidase	*Penicillium nomum*	1. Removal of glucose from eggs to be dried, to prevent fermentation; 2. Removal of oxygen from canned fruits, dried milk, and other products subject to oxygen spoilage
Invertase	*Saccaromyceturevisiae*	Preparation of soft-center candies (cordial cherries)

adjustments of pH, temperature, nutritive supplements and so on are made. In a pure-culture process, the mash is steam-sterilized, the entire fermentation tank sometimes being the autoclave. The inoculum, a *pure culture,* is added from a separate pure-culture apparatus. The fermentation proceeds. Some pressure may be maintained within the tank to prevent inward leakage of contamination and sometimes to maintain increased tension of special gases. After the proper time, the contents of the fermentor are drawn off for further processing, the fermentor is cleaned, and the process begins over again. Each fermentations is a discontinuous process divided into *batches.*

The Continuous-Growth Process : In continuous-growth processes, the substrate is fed into a container continuously at a fixed rate. The cells grow (or enzymes act) continuously as the material passes through the apparatus. The organisms and process are said to be in a *Steady state* or condition of *homeostasts.* The product or fully fermented mash is drawn off continuously. The engineering arrangements may be complex, permitting aeration, cooling or heating adjustment of pH or addition of nutrients continuously during the process. There must also be means of controlling rate of growth, phase of growth curve, and removal of dead organisms. The culture must remain pure and must not undergo any variation.

One may conceive of such a process as taking place in a long pipe (actually it may be a roatlng conical tank or series of connected tanks). At one end the prepared mash enters. It at once encounters the growing organisms. These act on the substrate as it flows through the system. At the stage at which the valuable product of the fermentation is at its maximum concentration, the fluid is drawn into receiving vessels for further processing (e.g., distillation). The animal alimentary tract may be thought of as a natural, continuous-growth process.

Submerged Aerobic Cultures : Many industrial processes, casually called "fermentations," are carried on by strictly aerobic micro–organisms: for example, production of penicillin by *Penicillium notatum,* a strictly aerobic mold. In older aerobic processes it was necessary to furnish large surfaces of culture media exposed to air. The limitations of space, difficulties from contamination, and expense of hand labour can well be imagined, though little expense for power equipment was necessary. Now it is common commercial practice to carry on such "fermentations" in closed tanks with *submerged cultures.* Aerobic conditions are maintained by constant agitation of the contents of the tank with an *impeller* and constant *aeration* by forcing sterilized air through a porous *diffuser.* The flow of air through the tank removes gases such as ammonia and carbon dioxide. In each sort of process very careful adjustments of O-R potentials, mechanical agitation, ratio of dissolved oxygen to other ingredients in the medium, and pH are necessary. This is one of the many fascinating and potentially very lucrative fields for research in industrial microbiology.

Industrial Ethyl Alcohol Manufacture

Much industrial ethyl alcohol is now made from by-products of *cracking* petroleum to make gasoline. However, the manufacture of ethyl alcohol from fermentation by yeasts is still an important industry. It serves to illustrate industrial fermentation processes in general. Crude molasses is often used as mash. It generally requires only to be diluted and the pH adjusted (usually with sulphuric acid) to 4.5. This pH is favourable to the yeast and unfavourable to many bacteria. A source of nitrogen such a ammonium sulphate or ammonium phosphate is usually added. The final solution is a richly nutrient carbohydrate culture medium of *mash.*

This is rather heavily inoculated with an aciduric and alcohol-resistant strain of yeast, the variety depending on the conditions under with the fermentations is to proceed and the exact end products desired. A good strain of *Saccharomyces cerevisiae* is commonly used. The inoculum comes from a large tank of carefully maintained pure culture, previously inoculated from a smaller seed tank, and the latter from a flask or tube of culture in the laboratory. At present stainless steel continuous-culture apparatus is available for maintaining constantly large amounts (many gallons) of *pure* cultures of inoculum.

Table 12.3 Source and uses of vitamins in the human body and vitamin deficiencies in humans.

Vitamin	*Food Source*	*Use in Body*	*Symtoms of Deficiency*
Thiamine (B_1)	Peas, beans, eggs, liver	Coenzyme use in Krebs cycle	Beriberi—breakdown of nerve cells, muscle and heart failure.
Riboflavin (B_2)	Milk, whole grain cereals, green vegetables, liver, eggs	Part of coenzyme used in electron transfer system (FAD)	Cracking of skin around the eyes and mouth; skin infections
Pyridoxine (B_6)	Most foods	Coenzyme used in synthesis of amino acids	Vitamin is so readily available that no deficiency disease has been noted.
Cyanocobalamin (B_{12})	Meats, dairy products	Used in red blood cell formation	Permicious anaemia–defective formation of red blood cells.

The maintenance of purity of the inoculum is a responsibility of the microbiologist, and woe betide him if some sporeformer, *Lactobacillus*, wild yeast or bacteriophage gets in and ruins 100,000 gallons of mash! The mash and all of the machinery are generally sterilized before the inoculation and then cooled. The microbiologist is kept busy at every stage of the process, making cultural and microscopic examinations of the water, mash and apparatus to detect and eliminate contamination.

In the batch process, much used for this purpose, fermentation in enormous tanks is allowed to continue for about 48 hours at a carefully controlled temperature of about 25°C until the yeast stops growing because of the concentration of alcohol and other products. Aeration with filtered air is used at first to promote rapid growth, but anaerobiosis is soon established to promote fermentation and alcohol accumulation, and prevent its oxidation to carbon dioxide and water.

The fermented mash contains the crude alcohol or *high wine*, as it is called. This is usually a mixture of ethyl alcohol and a small amount of glycerol with *fusel oil*. The last contains amyl, isoamyl, propyl, butyl and other alcohols with acetic, butyric and other acids, as well as various esters. The high wine is driven off from the mash or *beer* by heart, and further purified by fractional distillation, which is a problem in chemical engineering.

The Chemical reactions involved in the fermentation are complex; the principal stages follow the Embden-Meyerhof scheme. The overall reaction in the production of alcohol from glucose is:

$$C_6H_{12}O_6 \xrightarrow{\text{yeast}} 2C_2H_5OH + 2CO_2.$$

The chief constituents of fusel oil are probably derived from the action of the yeasts on amino acids in the mash. The large amounts of carbon dioxide evolved are purified and compressed in tanks or made into solid carbon dioxide. Part of this may be used for cooling the fermentation vats.

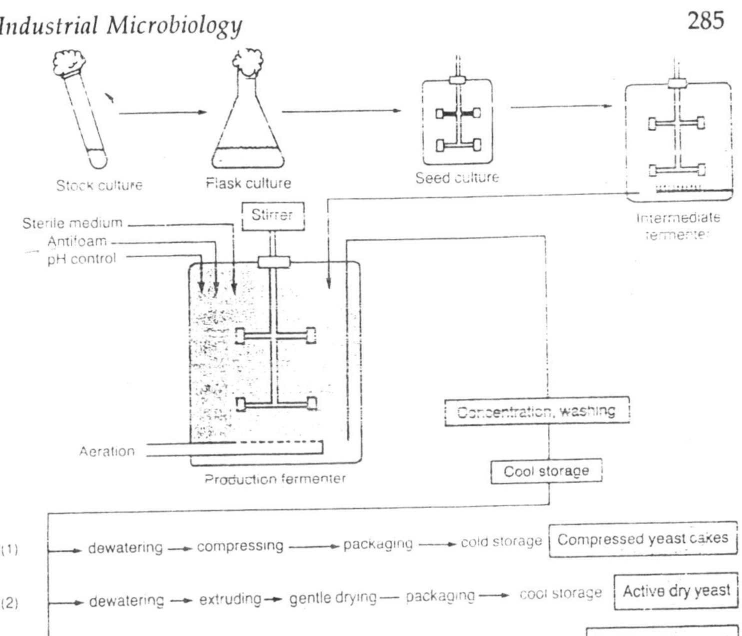

Fig. 12.2 The industrial production of *yeast, Saccharomyces cerevisiae.*

Alcoholic Beverages

Wine

Alcoholic beverages can be made by fermenting almost any grains, fruits, or vegetables that contain sugar. It was not uncommon for the early colonists in the United States to ferment pumpkins, potatoes, Indian corn, or carrots to make a beer. They fermented and distilled honey to make a rum substitute known as "old methaglin." Apple cider (unfiltered and non-pasteurized apple juice) was fermented to "hard cider." Hard cider was sometimes placed in wooden barrels and frozen in a lake during the winter to separate the water and concentrate the alcohol. This "Cold distilled" beverage was known as applejack and was very potent. Some prepared a mixture of applejack and gin in a drink known as "strip-and-go-naked" because of the strange behaviour it caused in those who drank only a few mugfuls. Other fermented fruits included elderberries, plums, peaches, currents, raspberries, and grapes. Today grapes are the primary fruit used in the production of wine. *Wine* is the fermented juice of grapes and other fruits with an alcohol concentration of between 9 and 21 per cent. Table or dinner wines contain no more than 14 per cent alcohol, while dessert and appetizer wines may contain up to 21 per cent alcohol. *Beer is* the fermented extract of grains such as barley or rice with an alcohol concentration of between 3 and 6 per cent. *Packaged liquors* are distilled fermentation products that have alcohol concentration as high as 85 per cent.

Grape wine may be made from any of 5000 varieties of the plant *Vitis vinifera* or 2000 varieties of other *Vitis* species. Each has a characteristic chemical contents that imparts a unique flavour, aroma (bouquet) and colour to the wine made from each variety. Because the chemical content of the grape is greatly affected by the soil and climate in which it is grown, grapes used for making good wines are only grown within certain wine-growing regions around the world.

Grapes contain a variety of sugars but the two most important are glucose and fructose. Glucose is fermented by yeast to carbon dioxide and alcohol, and the fructose gives a sweet taste to the wine. Since fructose tastes twice as sweet as glucose, grapes with a high percentage of this sugar are excellent for the production of sweet white wines. Also found are organic acids, free amino acids, proteins, pigments, and various minerals. The particular balance of these nutrients and the wine-making process affects the characteristics and quality of a wine. Grapes are picked from the vine at a time when they are most likely to have the desired balance of sugar and other compounds. They are quickly taken to the winery where they are mechanically crushed.

The stems are separated from this mixture of pulp, skins, and seeds called "must". The skins are also removed if white wine is to be made. Red and rose wines get their colour and flavour from pigments and tannic acid in the skins. "Must" is pumped to the fermenting vat made of hardwood or glass-lined concrete. Once inside the vat, sulphur dioxide or sulphurous acid is added to inhibit the growth of wild, undesirable strains of yeast that occur naturally on the grapes. It also interferes with natural enzymes that "brown" the wine, destroys grape skin cells aiding in the release of red pigments. Acidifies the must, and aids in clarifying the wine after fermentation. Pure cultures of *Saccharomyces cerevisiae* var, *ellipsoideus* are unaffected by the

Table—12.4 Taxonomy of wines

Wines With No Added Flavours		
Sparkling Wines (carbonated with and excees of CO_2) Champagne, Spumante, Sekt Pink champagne Sparkling burgundy and Cold Duck Sparkling muscat		
Still Wines (having no excess CO_2)		
Table wine (less than 14 per cent alcohol)		
White; Named by district Macon,	:	France : Chablis, Graves, Meursault, Montrachet, Pouilly-Fuisse, Sancerre Germany : Franken, Moselle (Piesport), Rheingan (Haiten-heim,) Rheiahessen
Named by variety	:	Chardonnay, Reisling, Chenin Blanc
Pink or Rose		Grenache, Cabernet Rose
Red; Named by district	:	France: Beaujolais, Bardeaux, Burgundy, Chateauneuf-du-Pape, Hermitage, Medoc, Pomerol, Rhone, Saint-Emilion Italy : Barolo Spain : Rioja
Named by variety	:	Cabernet Sauvignon, Pinot Noir, Zinfandel, Nebbiolo, Meriot
Dessert Wines (having more than 14 per cent alcohol) Sherry, dry to sweet Madeira Marshala Muscatel Port		
Wines with Added Flavours Herb-based Vermouth, dry Vermouth, sweet Dubonnet		
Fruit-based and fruit flavour- based Blackberry, peach, gooseberry, Cherry, applo winos		

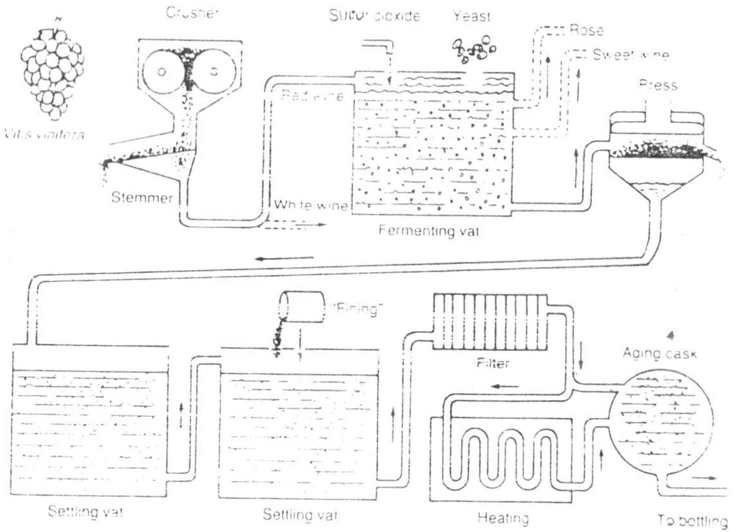

Fig. 12.3 This is only one method among many that might be used commercially to prepare wine.

sulphur dioxide and are added to the vat to ferment the must for a few days to a few weeks. Heat generated by the yeast is carried away by cooling coils in order to maintain a fermentation temperature below 85°F for red wines and below 60°F for white wines. An increasingly popular wine in the United States is rose. The production of rose wines begins as with any other red wine, but the skins are removed early in the initial fermentation to prevent the development of a dark red colour and heavier tannic acid concentration.

When this portion of the wine-making process is complete, the wine is pumped to a wine press to remove the pieces of skins, seeds, and other solids. The wine is present to a settling vat where the fermentation is allowed to continue, while smaller solids and yeast cells settle out. During this time, *lactobacillus* species convert strong malic acid to weak acids, thus reducing the sharp flavour of the wine. When the wine is clear and has matured, it is "racked," or drawn off, to another settling tank.

There the filtering process further clarifies the wine and reduces its oxygen content. Wines that lose their carbon dioxide during this process are called *still wines*. In many large wineries, clarification is added by the addition of "fining" agents such as clay, gelatin, or egg white which precipitates extremely small particles in the wine. After clarification and fining, it is pumped into oak barrels or casks for aging.

During the aging process, chemical reactions occur which develop the wine's flavour, aroma, and colour. The yeasty flavour is lost and red wines develop their typical rich red colour, while white wines become amber. The aging process usually requires about two years for red wines and varies from several months to about two years for white wines. The decision to bottle is made by the wine master and is very crucial in the production of a fine wine. However, the aging process does not stop after bottling. Red wines continue to develop and mature for years after bottling. Most wine masters and connoisseurs feel that red wines are their best five to ten years after bottling and white wines reach their peak after about two years.

Sparkling wines and champagnes are made in a unique way to capture and retain the natural effervescence of the carbon dioxide released during fermentation. The sparkling wine *Champagne* is only made in the Champagne region of France, while the uncapitalized name champagne is used as a generic name in the United States to describe any wine that contains an excess carbon dioxide level. When Champagne is made, the must is fermented and fined as with all wines. However a second fomentation is initiated after the wine is bottled by the addition of more yeast and sugar. These specially designed bottles are wired shut to prevent the gas from escaping and allowed to ferment neck-down on a special rack for a few weeks to several months—a process called tirade.

When the fermentation is complete, dead yeast, tartar, and other compounds settle on the inside of the bottle. This sediment must be removed without disturbing the carbonation or quality of the wine. This is done by "riddling," a special bottle-turning done by hand that causes the sediment to drop into the neck of the bottle. A talented riddler can riddle, or turn, about 30,000 bottles of Champagne a day. Each bottle must be turned once a day for about a month to sediment all the yeast into the neck. The bottles are then cooled almost to freezing, so that they can be turned upright and the corks removed along with the frozen sediment. It is at this time that sugar syrup is added to give the Campagne a more appealing taste. Four different types of Champagne are produced and vary according to the percentage of sugar added at this time. Brut is the driest (0.5-1.5 per cent sugar), sec has 2.5-4.5 percent sugar, demi-sec is more sweet (5 per cent), and doux the "sweetest" with 10 per cent added sugar. If riddling and tirage have been done well, the product is a clear, sparkling wine of high quality.

Beer

Another popular alcoholic beverage, beer, is believed to have been first produced by the Egyptians as the direct product of the fermentation of barley. As with wine the production of a fine beer requires the skillful blending of science and art. Several different types of beer are made throughout the world and vary in their colour, aroma, flavour, and alcohol content.

The brewing of beer begins with the selection of high quality barley and a process known as malting. *Malting* involves steeping (soaking) the grain in water of 65°C to initiate germination, or sprouting. During seed germination, natural enzymes convert the stored food material in the seed (endosperm) from starch to glucose. This increases the percentage of glucose available for yeast fermentation and reduces the amount of starch to a level that prevents the development of fusel oil during fermentation. The malted grain is then dried in kilns or rotatining drums at from 75° to 100°C. The actual temperature used for drying depends on the kind of malt used in the preparation of the brew. Lower temperatures are used for light colour malt and higher temperatures are used for dark malt. Care must also be taken to preserve the barley enzymes (amylase) needed to convert additional starch to sugar during the *mashing* process. In this process, the malt is cleaned, ground, weighed, and mixed with other cereal grains such as corn meal, soybean meal, corn grits, or corm flakes. These *adjuvants* help balance the protein and carbohydrate content to

Table—12.5 Type of beers and their characteristics

Type	*Fermentation*	*Yeast*	*Per Cent Alcohol by Weight*	*Characteristics*
Lager	Bottom	*Saccharomyces carlsbergensis*	3.8	Most popular beers in U.S. lightly hopped, fermented at low temperature.
Bock	Bottom	*S. carlsbergensis*	5.5	Heavy sugar, malt, and hops; few adjuvants; dark colour, a lager beer, fermented at low temperatures.
Pilsener	Bottom	*S. carlsbergensis*	3.8	A lager beer, light colour, low sugar, medium hopped.
Malt liquor	Bottom	*S. carlsbergensis*	4.2	Lightly hopped, heavily malted.
Light beers	Bottom	*S. carlsbergensis*	3.8	A lager beer, double fermented to reduce sugar and diluted to reduce alcohol content.
Ale	Top	*S. cerevisiae*	4.7	Heavily hopped, high sugar, fermented at high temperatures.
Porter	Top	*S. cerevisiae*	4.7	Dark ale, sweet, lightly hopped, heavily malted.
Stout	Top	*S. cerevisiae*	4.7	Strong porter, heavily malted and hopped.

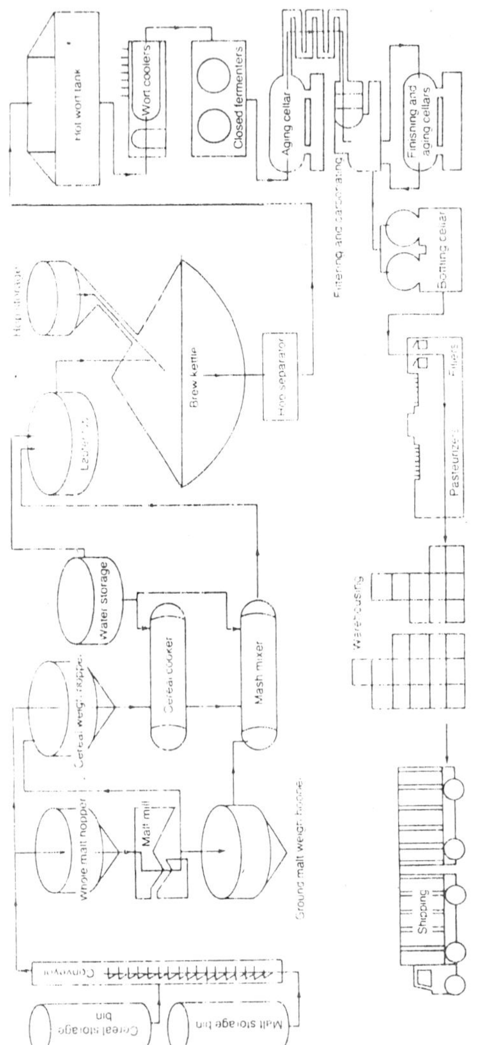

Fig. 12.4: The brewing process. (Courtesy of the G. Heileman Brewing Company).

ensure a more flavourful product. The mixture is sent to a lauter tub, where it is purged with hot water to extract the nutrients from the mixed grains. The leftover of "spent" grain is dried and sold for used as animal feed. The protein content of this grain is very high, thus it is an excellent supplement to cattle and hog feed. The fluid extract, known as *wort* (pronounced "*wert*"), has the colour of beer and is pumped to the brew kettle where it is cooked for 1½ to 2½ hours. While in the brew kettle, flowers of the female hop plant are added to give beer its characteristic aroma and inhibit the growth of contaminating bacteria that could later interfere with fermentation or cause defects. The most important contaminating bacteria are *Lactobacillus pastorianus, Pediococcus cerevisiae, Flavobacterium proteus*, and *E. coli.*

After cooking in the brew kettle, the wort is passed through a hop separator that not only removes the hops but also extracts undesirable resins and residues. The clarified wort is then cooled and pumped into a closed fermenter where an active culture of *Saccharomyces* is "pitched", or inoculated, into the vat. Various species and strains of *Saccharomyces* are available for brewing. They are chosen for their unique genetic ability to ferment the wort at desirable rate and determine the character of beer. Brewery yeast are classified as either bottom-fermenting yeast (*e.g., S. carlsbergensis*) or top-fermenting yeast (*e.g., S. cerevisiae*). Bottom fermenters are used for the production of lager beer (the more common American beers) while top fermenters are used in the manufacture of ale. The fermentation usually lasts between seven to ten days (depending on whether it is top or bottom fermented) and is maintained at a temperature of about 40°C. As the fermentation proceeds, alcohol, acetic acid, glycerol, and carbon dioxide are formed. The carbon dioxide is piped from the fermenter, compressed, and stored. It is later returned to the beer during the carbonating process. When the fermentation is complete, the yeast population has increased six times and settled to the bottom of the tank. A portion of this yeast is washed and reused for other fermentations, while the excess is dried and added to poultry and hog feed. These yeast may also be used in the fermentation of corn or sugar beets to produce alcohol for gasohol. The beer is pumped from the fermenter for aging (lagering), filtering, carbonating and packaging. Beer in bottles and cans is pasteurized to lengthen its shelf life, but beer in kegs is not pasteurized and must be refrigerated.

In order to maintain production of a high quality beer, quality control measures are utilized throughout the brewing process. Equipment is cleaned and disinfected on a regular basis. Laboratory personnel maintain pure, active cultures of yeast, and a special effort is made to select for reuse only those yeast cells that will develop the highest quality beer. The final product is regularly checked to identify the presence of contaminating bacteria that may be responsible for beer defects.

Distilled Alcohol as Beverage and Fuel

Alcohol produced during fermentation does not accumulate in the culture above a level of about 18 to 21 per cent, since it is toxic to the yeast producing it. For this reason, a simple fermentation does not result in a beverage with an alcohol concentration greater than 18 to 21 per cent. (The exact percentage depends on the strain of *Saccharomyces* used). In order to produce an alcoholic beverage with a higher concentration, *i.e.*, hard *liquor,* the beverage must be distilled from the original fermentation. *Distillation* is a heating process that drives alcohol vapour and some of its dissolved components from water and condenses it to a liquid that can then be purified, filtered, and clarified for consumption. Among the first alcoholic beverages distilled into liquors were wines (mentioned by Aristotle) and beer (described in early European writings). Today a variety of liquors are made by distilling alcoholic beverages fermented from many types of starchbased grains, and sugary fruits and berries. Each beverage is made from special starting material, using fermentation techniques unique to each type of liquor. For example, Scotch whiskey is made from barley malt that has been kiln dried at a high temperature over a peat moss fire. This gives the whiskey its unique flavour. Bourbon is made from barley malt that has between 51 and 80 per cent added corn, and some rye is always added. Canadian whiskeys are fermented mainly from rye. The final concentration of alcohol in a liquor (given in "proof" number) is controlled by the distillation process and dilution with water. Packaged liquors such as vodka and gin are usually sold with a final proof of 80 or 90, but may be as high as 170 proof. The proof number is twice the alcohol concentration, *i.e.*, 170 proof vodka has an alcohol concentration of 85 per cent. Grain alcohol has a higher final proof, and its sale is regulated by state liquor departments. In most cases, this alcohol can only be purchased from a druggist and only with a valid letter of authorization describing its intended use.

In recent years the energy shortage has rekindled interest in ethyl alcohol as a liquid fuel to be used in automobiles, home heating systems and elsewhere. The idea is not really old because alcohol lamps, stoves, and many other appliances have been fueled with clean-burning alcohol for hundreds of years. Ethyl alcohol may be used as a liquid fuel in combination with gasoline as gasohol or as the more purified grain alcohol (190 proof). As an engine fuel, grain alcohol has several advantageous characteristics. It can be burned in an engine with only minor modification and with the efficiency of a high octane gasoline that prevent engine "knock".

There are four general types of distilled liquor: brandy, from fermented fruit juices; rum, from fermented molasses, whiskey, from fermented masses made with single types of the grains; neutral spirits, from fermented mash of mixed grains. In making whiskey and neutral spirits the grain, mixed with water, is autoclaved, cooled, diluted, and 1 per cent barley *malt* (aqueous extract or sprouted barley; see next section on beer) is added to hydrolyze the starch and proteins of the grain. The "mashing," or hydrolysis, proceeds in a special tank at about 65°C, for about 30 minutes. The mash is then pumped to the fermentation tanks. Here, as in beermaking, it is heavily inoculated with a starter of selected year, which has been cultivated in a mash previously made somewhat acid (pH 4.0) with lactobacilli. Fermentation is complete in about 72 hours, as in industrial alcohol production. The mash is then removed to the distillery and the ethanol, with various by-products, is recovered.

Beer

This time-honoured and popular beverage is one of the class of malt liquors: stout, porter, ale and others. In preparing beer, grain, usually barley, is kept moist for two or three days to induce sprouting, or *malting*. Amylase enzymes that are released in the malt grains during the sprouting process hydrolyze the starches of the grains to simple sugars, mainly maltose and dextrins. Malting (Ger. *malz* = *to* soften) is necessary since brewer's yeast does not produce amylase and therefore cannot directly attack the starch of the grain. At the same time proteases in the malt grains convert proteins in the grains and flour to soluble nitrogenous foods.

The sprouts are removed mechanically and the *malt gains* are dried. They are later crushed and soaked, or *mashed*, in warm water. The aqueous extract of these malt grains and flour, prepared at just the time when there are maximum amounts of maltose, dextrins and protein derivatives, constitutes a rich nutrient medium. It is called *beer wort*. The beer wort is now drained off and heated to kill contaminating micro–organisms. Hops are added for additional antibacterial (stabilizing) effect, colour, flavour and aroma.

After cooling, a large inoculum of pure culture of *Saccharomycs cerevisiae* (brewer's yeast or *"barm"*) is added to the wort. This is called *"pitching"*. Rival brewers maintain very special strains of yeast for this process. The inoculated wort is aerated at first to stimulated *rapid growth* of yeast; anaerobic conditions prevail later on to favour *fermentation,* when carbon dioxide and 3 to 6 per cent ethanol are produced. After fermentation is compete, the beer is clarified ("chill-proofed"); and pasteurized and otherwise processed and aged (lagered).

Unless scrupulous care is taken, many contaminants (*Pediococcus, Lactobacillus*) will grow vigorously in beer wort, producing buttery flavour, turbidity and "off" flavours. It was in the study of such spoilages, or "diseases," of beer and wines that Pasture first became famours and developed *pasteurization* to prevent them. He was one of the first industrial microbiologists.

Wine

The term wine is broadly used to include and properly fermented juice of ripe fruits, or extracts of certain vegetable products such as dandelions and palm shoots. The juices or extracts contain glucose and fructose in concentrations of from 12 to 13 per cent. Fermentation of these sugars by various species of yeasts produces carbon dioxide and ethanol up to concentrations of 7 to 15 per cent, the alcoholic content depending on the kind of juice and yeast involved and the conditions of fermentation. In Europe the fermentation is produced mainly by *wild* yeast, *i.e.,* those brought to the fruits (largely by insects) from soil or other fruits. Yeasts similar to the species called *Saccharomyces ellipsoidues* are common in such wines.

Although other organisms are usually present, the yeasts soon predominate in the fermenting juice under suitable conditions. Tartaric, malic and other acids, as well as tannin and other substances, including added sulphur dioxide in commercial wine, tend to inhibit growth of many undesirable organisms in the juices.

Even though practices may differ in different wineries, basically they are similar. Commonly in modern American commercial practice, sterilized fruit juices are inoculated with a pure culture of a desirable species of yeast. The preparation and maintenance of the yeast inocula are the special tasks of the microbiologist.

The inoculated juice is, as in beer-making, at first aerated to promote active and preemptive growth of yeast. Were this to continue, only carbon dioxide and water and massive growth of yeast cell would result. As soon as a good growth of yeast has occurred, the aeration is stopped and the fermentation proceeds anaerobically, so that ethanol, in concentrations of from 7 to 15 per cent (vol.) is produced. The new wine is placed in large casks to settle, clarify and age.

Spoilage by alcohol-oxidizing species of *Acetobcter,* molds and other *aerobic* micro–organisms may occur if conditions are not anaerobic and the reaction not sufficiently acid.

Production of Butanol

The production of butanol is outlined as an example of an industrial fer nentation based on a species of bacterium.

As is true of a industrial ethanol production, much butanol is now derived as a by-product of petroleum "cracking." However, the biological process is still used to some extent and illustrates important microbiological principles. There are numerous species of *Clostidium* that ferment carbohydrates with the production of butyl alcohol and other materials of value in drugs, paints, synthetic rubber, explosives and plastics. Some species produce isopropyl alcohol and acetone as well. Important among these organisms are *Cl. acetobutylicum* and *Cl. felsineum.* The name of another species suggests its potentialities as an industrial agent: *Clostridium amylosaccharobutylpropylicum!*

Many wastes are rich in fermentable carbohydrates, *e.g.,* cannery refuse. Complete sterilization of all apparatus is essential. Conditions cannot be kept as acid (and antibacterial) as they are in yeast fermentations because *Clostridium* has its optimum pH near 7.2. Particularly troublesome contaminants are species of *Lactobacillus.* An organism called *B. volutans,* a grampositive, non spore forming rod (possibly a species of *Lactobacillus?*), is also especially dangerous.

Fermentation proceeds anaerobically for about three days. Normally butyl alcohol, acetone and ethyl alcohol, with carbon dioxide and hydrogen in large amounts, predominate when *Cl. acetobutylicum* acts in a mash rich in glucose. Other substances may occur in smaller amounts. Riboflavin (a vitamin of the B complex) in a valuable constituent of the residue after distillation of the fermental mash. Butyl and isopropyl (rubbing) alcohols are important among the volatile fermentation products of a related specics, *Cl. butylicum.*

The successive reactions in the production of butanol and various side-products from glucose are as follows:

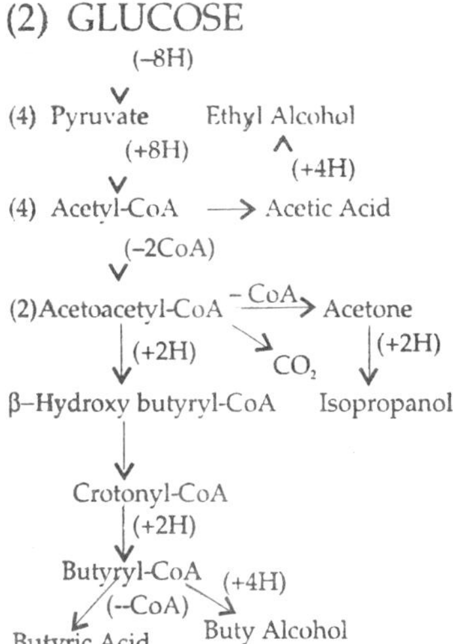

Lactic acid is commonly produced from lactose or glucose by species of homo-fermentative lactic acid bacteria, notably *Streptococcus lactis* or *Lactobacillus delbrueckii;*

$$\underset{\text{Lactose}}{C_{12}H_{12}O_{11}} + H_2O \longrightarrow \underset{\text{Glucose and Glactose}}{2C_6H_{12}O_6} \longrightarrow$$

$$\underset{\text{Pyruvate}}{4C_3H_6O_3} \xrightarrow{4H_2} \underset{\text{Lactic acid}}{4CH_3.CHOH.\ COOH}$$

Heterofermentative species of lactobacilli like *L. buchneri* purpose lactic acid and several by-products:

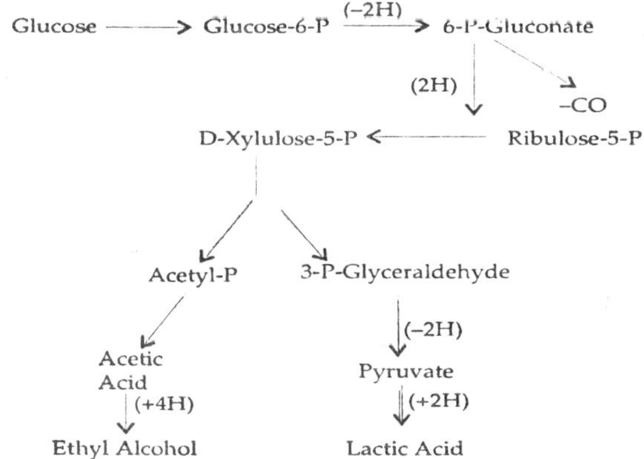

Production of Vinegar

Acetic acid is almost entirely responsible for the sour taste of vinegar. Indeed, a slightly sweetened, three per cent, aqueous solution of acetic acid makes a reasonable substitute for vinegar.

The occurrence of vinegar in fermented fruit juices was known to the ancients, although they had not knowledge of its cause. The bacteria involved were called *Mycoderma aceti* in 1862 by Pasteur.

The acid of natural vinegar is derived from alcohol by the oxidative action of bacteria of the Family Pseudomonadaceae (Genus *Acetobacter*). Pleasant flavours of natural vinegar are given by traces of various esters like ethyl acetate, and by alcohol, sugars, glycerin and volatile oils produced in small amounts by microbial action. Flavours are also derived from the fermented fruit juice, malt, or other alcohol liquor (wine, beer, hard cider) from with the vinegar was made.

In commercial vinegar-making by biological methods, preliminary fermentation of fruit juices to produce the necessary alcohol is often carried out by means of *Saccharomyces cerevisiae* (brewers' yeast). The *Acetobacter* then utilize the alcohol as a source of energy, oxidizing it to acetic acid in the presence of air. They utilize other substances in fermented liquor as foods. The alcoholic fluid is aerated as much as possible by various devices. In vinegar" generators" the alcoholic liquor trickles over the surface of aerated shavings, coke, gravel or other finely divided material inoculated with *Acetobacter.* Such an arrangement is called a *two-phase continuous process;* one phase is the down-trickling alcoholic liquor, the other phase is the column of coke or other material covered with growth of *Acetobacter.*

Genus Acctobacter

These are nonsporeforming, polar or peritrichous flagellate, gram-negative rods about 0.5 by 8.0 μ, although species vary in size. Branching involution forms and large swollen cells frequently occur, especially in *mother-of-vinegar,* the gummy or slimy growth-phase of the organisms sometimes seen in natural vinegar or sour cider. Various species are found in souring fruits and vegetables. A species of historical interest is *Acetobactor*

(Mycoderma) aceti, originally used by Pasture to demonstrate the biological nature of vinegar formation. In practice, several species of *Acetobacter* usually act jointly. The alcoholic and acidic nature of the process suppresses most contaminants. The overall reactions probably are as follow:

$C_2H_5OH+½O_2$ (*Alcohol*) ⟶ $CH_3CHO—H_2O$ (*Acetaldehyde*)

$CH_3CHO+½O_2$ ⟶ CH_3CHOOH (*Acetic acid*)

In a generator the rapid oxidation of alcohol by the organisms produces so much heat that careful control of the internal temperature by cooling coils is necessary.

FOODS FROM WASTE

In the paper-pulp industry, wood chips are cooked for 6 to 18 hours at 60°C in solution of calcium bisulphite with free sulphur dioxide. The waste *sulphite liquor*, after the cooking process and removal of the wood fibers for papers, contains much valuable wood sugar (largely *xylose*) and other extractives. These form a good nutrient for asporogenous yeast or *Torula*.

The nutrients in such a medium may be turned into masses of yeast by adjustment of pH to about 5.0, removal of SO_2, aeration, additioin of nitrogen and phosphorus as $(NH_4)2HPO_4$ and NH_4OH, and inoculation with *Torulopsis utilis*. *Aerobic* growth is induced in aerated vats so that alcohol is not produced. The separation, drying and pressing of the resulting yeast growth are mechanical details. Yields of up to 50 per cent of the total reducing sugar consumed, in terms of dry torula, are obtainable. (At present, attempts are being made to cultivate yeasts on petroleum wastes). The yeast cells are rich in proteins, facts and vitamins. These are fed to stock or poultry and thus turned into meat and dairy products. Surely the transformation of a knotty old pine slab into a succulent pork chop or a fired egg is modern magic!

Amino Acid Production

Not only may yeasts serve as foods themselves, but some species, especially *Saccharomyces cerevisiae* and *Torula utilis*, while growing can synthesize large amounts of various amino acids.

e.g., L-lysine. This is an expensive amino acid widely used to "fortify" many familiar foodstruffs. It is requisite that the growth medium for the yeast contain L-adipic acid or its derivatives. The yeasts use the adipic acid derivatives as *precursors* (*i.e.*, as the molecular raw material) for their synthesis of L-lysine.

Hydrocarbons for Protein

Various hydrocarbon (petroleum) wastes are metabolizable by certain yeasts, eucaryotic fungi and also by some bacteria, *e.g.*, *Bacillus* spp., especially thermophilic species. One difficulty (or possibly a source of profit?) is the production of large amounts of heat by biooxidation of hydrocarbons. The proteins produced in the process are of high nuritive value and the method is potentially profitable.

Table—12.6 Examples of distilled alcoholic beverages

Name	*Starch Base*
Scotch	Barley
Bourbon	corn, rye, barley
Rye	Rye and barley
Vodka	Potato
Irish whiskey	Rye, wheat, barley
Gin	Grins or potatoes with juniper berries
Arrak	Rice
Cognac	Grape
Brandy	Grape
Sloe Gin	Plum
Tequila	Agave (century plant)
Rum	Sugar cane
Applejack	Apples
Toddy	Coconut milk
Framboise	Raspberries
Liqueur and cordial	Mixtures of distilled alcoholic beverages, herbs, sugar, and flavourings

Microbial Enzymes in Industry

The enzyme industry is large and growing as more information relating to microbial metabolism becomes available through research. The industrial microbiologist applies this information in an attempt to increase the yield of a profitable product. Because these applications may be patented, much of the detailed information surrounding the actual growth of microbes and harvesting of enzymes is kept secret.

Enzymes are protein catalysts that speed the rate of specific chemical reactions and are produced through a complex sequence of protein synthesis reactions inside a living cell. The exact nature of an enzyme and its action is determined by the sequence of nucleotides found on the DNA of the cell in which it is manufactured. This fact makes it possible to select a particular genetic type of microbe for the production of an enzyme. The highly specific nature of enzyme action and the speed with which enzyme increase reactions make them excellent chemicals in many industrial, commercial, and medical processes. Microbes are excellent sources of enzyme, since they are easily grown in large aerated and submerged cultures, and produce enzymes on a regular basis. While in culture, bacteria, fungi, and filamentous bacteria synthesize and excrete *exoenzymes* that accumulate in the surrounding medium. Other enzymes are called *endoenzymes* since they are synthesized and utilized by the microbe inside the cell. To harvest these enzymes, the cultured cells must be separated, washed, and fragmented before the enzyme is isolated and purified.

1. How does distillation differ from fermentation?
2. What is "spent" grain and what advantages are there in using it as animal feed?
3. What are the exoenzyme and endoenzyme?

Enzymes are highly specific in their actions and must chemically "fit" their substrate in order to function properly. Therefore, it is essential that they be produced and utilized in an environment that does not distort their shape or interfere with their action. One of the most widely used proteolytic enzymes that is frequently misused is found in household enzyme pre-soaks or as they are sometimes called "enzyme detergents." These enzymes are produced from cultures of the bacterium *Bacillus subtilis* and are able to hydrolyze many different proteins found in stains caused by chocolate, blood, and many foods. However, many people do not gain the maximum advantage from these products because they do not follow the product's directions, which are based on fundamental enzyme chemistry. These products should not be used in extremely hot water, because the high temperatures will denature the enzymes and prevent them from combining with substrate stains. Cold water slows their action and requires longer contact between the enzyme and substrate for effective stain-removal. By rubbing the enzyme directly into the stain, more effective and rapid contact is made and hydrolysis of the stain is more efficient. Since enzymes also lose their effectiveness in environments which contain inhibitors, they should not be used in certain types of "hard" water (water containing inhibiting minerals) or along with detergents such as the dodecyl-benesulphonate group, which includes most granular laundry detergents. Caution should also be taken with enzyme pre-soak products, since many people find them to be allergens. Wearing clothes cleaned in these products may cause allergic skin reactions Table 12.7 lists several other microbially produced enzymes used in many industrial, commercial, and medical processes and products. Their use must be carefully controlled. While both bacteria and fungi are used in the production of enzymes, it is the fungi that are primarily responsible for the production of large amounts of industrially important organic acids. They are used to produce resins which are, in turn, used for the manufacture of such items as plastics, varnishes, electrical non-conductors and medicines. Other organic acids are used for the production of such products as perfumes, oil additives, and textile dyes.

Vitamins and Amino Acids

Any molecule that is required in minute amounts for the efficient operation of an organism and must be supplied from the environment since it is unable to be synthesized by that organism is known as a growth factor. Two of the most important classes of growth factors are the amino acids and

Table—12.7 Some industrially produced microbial enzymes and their uses.

Enzyme	*Microbe Cultured (Genus)*	*Uses*
Pectinase	*Aspergillus*	Separate fibers of the flax plant that may be used to make linen; same process used to make hemp rope. Used in green coffee processing. Used to clarify fruit juices.
Protease	*Aspergillus* *Bacillus* *Streptomyces*	Meat tenderizer; digestive aid removes gelatine from photographic film for silver recovery; used in enzyme pre-soaks and detergents.
Amylase	*bacillus*	Clarifies syrups glucose production; softens dough in commercial bread making; removes starch and sizing from textiles; ingested as digestive aid, wallpaper remover.
Collagenase	*Bacillus* *Clostridium*	Digests necrotic tissue from burns.
Lipase	*Rhizopus*	Digestive aid.
Rennin	*Mucor*	Curd formation in cheesemaking.
Cellulase	*Trichoderma* *Streptomyces*	Digestive aid. "Sweating" of animal hides to remove hair before tanning to leather.
Invertase	*Saccharinyies*	Soft-centered candies.
Urease	*Fungal*	Urea determination.
Lactase	*Lactobacillus*	Whole milk concentrates; lactose removal from milk for lactose-intolerant persons.
Streptokinase and streptodorase	*Streptococcus*	Digests necrotic tissue from burns.

vitamins. Although humans lack the ability to produce some of these compounds, many microbes are able to synthesize them. Therefore, what is a growth factor to a human may be only a normal metabolic by-product to these microbes. Despite the fact that many microbes have the ability to synthesize these complex compounds, few produce excess amounts since they are essential to the normal metabolism of the microbe and are, in most cases, quickly utilized. Microbes selected for the industrial production of human vitamins and amino acids are defective mutants that

Table—12.8 Microbially produced organic acids.

Acid	*Microbe (Genus)*	*Uses*
Lactice acid	*Lactobacillus*	Food preservative; in baking powder, as calcium lactate for food supplement for calcium deficiencies; bone development in pregnancies; used to make plastics.
Gluconic acid	*Aspergillus* *Penicillium*	Washing and softening agent; leavening agent; textile printing; As calcium gluconate for calcium deficiencies and in chicken feed to harden egg shells.
Citric acid	*Aspergillus*	In foods and beverages for taste and preservation.
Itaconic acid	*Aspergillus*	In manufacture of acrylic resins; plastic products; oil additives; dental resins.
Gibberellic acid	*Gibberella* *Fusarium*	Plant growth and flowering stimulant.
Kojic acid	*Aspergillus*	Analytical reagent; insecticide.
Fumaric acid	*Rhizopus*	used as wetting agent and in manufacture of resins.
Gallic acid	*Aspergillus*	Ink and dye manufacture; skin disease cream.
Ustilagic acid	*Ustilago*	Used in "musks" for the perfume fume industry.

have been isolated from wild cultures using special methods. These mutants lack the ability to use all the amino acids and vitamins they produce. The compounds accumulate inside the cells and are then released into the surrounding culture medium where they can be isolated and purified.

One of the most widely used amino acids produced industrially is glutamic acid. This amino acid is converted to the familiar compound monosodium glutamate (MSG) and added to many foods as a flavour enhancer. It is sold under the trade names Accent, Glutavene and Zest. MSG is added to many commercially prepared foods including baby foods, canned soups, meats, seafoods, and some soy sauces. MSG has been identified as a cause of brain damage; acute degenerative lesions have occurred in the retina of infant mice because of their age-dependent ability to absorb this compound. Although as yet unconfirmed, many researchers suspect that MSG may have a comparable ability to harm human infants. Since this food additive confers no significant benefit and, as yet, no confirmed harm to human infants, the Food and Drug Administration cautiously permits its unrestricted use, except for foods specifically designed for infants, However, many researchers question the safety of this chemical and continue to discourage its widespread use, especially in light of the fact that MSG has been confirmed to cause harm to adults. In 1968, it was discovered that this additive was responsible for Chinese restaurant syndrome, a disease characterized by headache, burning sensations facial pressure, and chest pain. These symptoms occur in adult after eating Chinese food prepared with high concentrations of MSG to intensify their flavours. In addition to glutamic acid, industrial microbiologists have produced other amino acids. Table 12.9 lists several these along with the most frequently produced vitamins, B_2 and B_{12}.

Table—12.9 Vitamins and amino acids synthesized by microbes

Product	*Microbes*	*Uses*
Amino acids		
Gutamic acid	*Micrococcus* spp. *Pseudomonas* spp. and *E. coli*	Flavour enhancer in foods as MSG (monosodium glutamate).
Lysine	*E. coli* and *Enterobacter* spp.	Food supplement, aids in preventing protein deficiency disease, kwashiorkor.
Threonine Methionine Tryptophan		
Vitamins		
B_2 (ribo-flavin)	*Ashbya gossypii*	Food supplement prevents cracking of skin, aids in preventing infections.
B_{12} (coba-lamin)	*Streptomyces* spp. *Pseudomonas* spp. *Bacillus* spp.	Food supplement, prevents pernicious anemia-defective formation of red blood cells.
Mixed Vitamins and amino acids	*Saccharomyces cerevisiae* and *Torula utilis*	Food supplement.

The greatest variety of human vitamins are manufactured by the yeasts *Saccharomyces cerevisiae* and *torula utilis*. These are sold in the form of compressed ironized yeast tablets and contain a mixture of many vitamins. The purified forms of vitamins are produced from cultures of bacteria such as *Streptomyces, Bacillus* and *Propionibacterium*. One of the most important is vitamins B_{12} (cobalamin) which is produced in excess by *Streptomyces* growing in glucose or corn-steep broth with traces of cobalt. Vitamin B_2 (riboflavin) is produced from cultures of the fungus *Ashbya gossypii*, Although this fungus is a cotton plant pathogen, its vitamin product is one of the most widely used vitamin supplements in the commercial food industry.

Antibiotics and Steroids

Antibiotics are unique molecules produced by micro-organisms and they are able to destroy or inhibit the growth of other microbes. The industrial growth of large, submerged and aerated cultures of fungi, filamentous bacteria, and true bacteria marked a major break-through in the ability of the medical profession to treat and prevent infectious diseases. By culturing antibiotic producing microbes in these large fermentation vessels, each cell is able to come in contact with nutrients essential to the production of the drug and maximize its output. The search for and detection of antibiotic-producing microbes relies on samples taken from water, soil, and other materials from many parts of the world. In one method, microbes suspected of being antibiotic producers are isolated from mixed samples and plated on nutrient media for growth in pure culture. After lawns of growth are formed on the surfaces of nutrient agar plates, small plugs are removed and placed on test plates inoculated with specific strains of *Staphylococcus aureus* or other pathogens. If any of the experimental microbes produce an antibiotic that is capable of destroying or inhibiting the pathogen, zones of clearing will form around the plug. These zones resemble those produced in a culture and susceptibility test performed in a medical laboratory. The microbes producing antibiotic are then grown in pure culture for further research to explore the nature of their metabolism and the likelihood that they will produce enough antibiotic to make them industrially profitable.

From agar slants the microbes are transferred to larger flasks to increase the number of cells. These are shaken continuously to ensure optimum accretion and growth. The shaker flask cultures serve as inocula for "bazookas" (transfer culture carriers) that are larger culture vessels used to increase the amount of inoculum to 10 per cent of the fermenter volume. Once inside the fermenter, the microbes grow, reproduce, and release significant amounts of the antibiotic. When the maximum amount of antibiotic has been produced, the solid components of the culture (*i.e.*, cells and other debris) are separated from the medium in a large rotary filter. The fermentation broth is then passed through a solvent extraction process that further separates the drug from unwanted chemicals. It is then transferred to an ionexchange chromatography apparatus that contains resins which adsorb the antibiotic to their surfaces. This process separates the drug from the remaining impurities before the antibiotic is crystallized under rigidly controlled conditions. The drug is then packaged in the form of capsules, tablets, liquids, or injectables. However, an alternative course can be taken to chemically modify the pure drug to other forms. This is done primarily to avoid the problem of drug resistance. As a result, entire families of antibiotics have produced from the fundamental compounds. The brief description of this process in no way reflects the years of intensive work needed to reach the point at which a microbe may be cultured for its valuable, government-approved products. In some cases, 10,000 or more species of strains of microbes have to be investigated before one suitable for culturing is identified, and years of research and field testing are necessary before the drug is made available to the public. It is also important to realize that of the over 2000 antibiotics described in research literature, only a few are of economic and medical value. The rest are far too toxic or ineffective to be used in the treatment of infectious diseases. Table 12.10 lists many of the microbes involved in antibiotic production, their products, and their spectrum of action.

The last medically important group of compounds the steroids, are produced by an industrial fermentation process known as bioconversion. *Bioconversion* is the use of microbial fermentation to enzymatically change one compound to another

Table 12.10 Antibiotics of clinical importance

Generic Name	*Source*	*Antimicrobial Spectrum and Some Important Properties*
Amphotericin B	*Streptomyces nodosus*	Fungi
Bacitracin	*Bacillus subtilis*	Gram-positive bacteria
Cephalosporins	*Cephalosporium* spp. and *Emericellopsis* spp.	Generally penicillinase-resistant but inactivated by cephalosporinase
Cephalexin	Semisynthetic	Gram-positive and some Gram-negative bacteria orally absorbed
Cephaloglycin	Semisynthetic	Like cephalexin; orally absorbed
Cephaloridine	Semisynthetic	Like cephalexin; not orally absorbed
Cephalothin	Semisynthetic	Like cephalexin; not orally absorbed
Chloramphenicol	*Streptomyces venezuelae;* synthesis	Gram-positive and Gram-negaive bacteria; rickettsias; *Entamoeba*
Colistin	*Aerobacillus colistinus*	Primarily Gram-netative bacteria
Erythromycin	*Streptomyces erythreus*	Gram-positive and a few Gram-negative bacteria; some protozoa
Gentamycin	*Micromonospora* spp.	Gram positive and Gram-negative bacteria
Griseofulvin	*Penicillium* spp. synthesis	Fungi
Kanamycin	*Streptomyces kanamyceticus*	Gram-positive and Gram-negative bacteria; some protozoa
Lincomycin	*Streptomyces lincolnensis*	Gram-positive bacteria
Neomycin	*Streptomyces fradiae* and other Steptomyces spp.	Gram-positive and Gram-negative bacteria; myco-bacteria
Novobiocin	*Streptomyces*	Primarily Gram-positive
Pencillins	spp.	Bacteria
Ampicillin	Semisynthetic	Gram-positive and Gram-negative

(Contd...)

Table 12.10 (Contd...)

Generic Name	*Source*	*Antimicrobial Spectrum and Some Important Properties*
		bacteria; penicillinase-sensitive; acid stable
Carbenicillin	Semisynthetic	Gram-positive and Gram-negative bacteria including *Pseudomonas* strains: penicillinase-sensitive; not orally absorbed
Cloxacillin	Semisynthetic	Gram-positive bacteria; penicillinase-resistan; acid stable
Dieloxacillin	Semisynthetic	Like cloxacillin
Methicillin	Semisynthetic	Gram-positive bacteria; penicillinase-resistant; acid labile
Nafcillin	Semisynthetic	Like cloxacillin
Oxacillin	Semisynthetic	Like cloxacillin
Penicillin G	*Penicillium* spp. *Aspergilbus* spp.	Gram-positive bacteria; penicillinase-sensitive acid labile
Penicillin V	biosynthetic	Gram-positive bacteria; penicillinase-sensitive acid stable
Polymyxin	*bacillus polymomyxa*	Primarily Gram-negative bacteria
Rifamycin	*Nocardia mediterranea*	Primarily Gram-positive bacteria and mycobacteria
Rifampin	Semisynthetic	Gram-positive and Gram-negative bacteria; mycobacteria; orally absorbed
Spectinomycin	*Streptomyces* spp.	Gram-positive and Gram-negative bacteria, specifically *Neisseria gonorrhoeae*
Streptomycin	*Streptomyces* spp.	Gram-positive and Gram-negative bacteria; *Mycobacterium tuberculosis*
Tetracyclines		Gram-positive and Gram-negative bacteria; rickettsias; coccidia; amoebae and balanticia; mycoplasmas
Chlortetracycline	*Streptomyces aureofaciens*	
Demeclocycline	*s. aureofaciens* mutant	

(Contd...)

Table 12.10 (Contd...)

Generic Name	*Source*	*Antimicrobial Spectrum and Some Important Properties*
Methacycline	Semisynthetic	
Minocycline	Semisynthetic	Also inhibits staphylococci resistant to other tetracyclines
Oxytetracycline	*Streptomyces* rimosus	
Tetracycline	Catalytic bydrogenation of chlortetra-cycline; *S. aureofaciens* mutant	
Tyrothricin	*Bacillus brevis*	Gram-positive bacteria
Vancomycin	*Streptomyces orientalis*	Gram-positive bacteria

biologically active form, in this complex series of fermentations, biologically inactive steroid compounds from plants are used as starting materials. A specific microbe is selected for its genetic ability to produce an enzyme that specifically alters the compound to another form. A series of several fermentations using different microbes is required to enzymatically change one compound into an other. The final product is a tailor-made molecule with the desired biological activity. Some of the medically important compounds produced by bioconversion include bile salts, progesterone, testosterone, vitamin D, desoxycholate, cholesterol, estradiol, and cortisone. Among the more important genera of microbes used for bioconversions are *Rhizopus, Streptomyces, Aspergillus Corynebacterium* and *Curvularia*. The complex fermentations required to convert one of the more basic compounds, progesterone, into a variety of other important molecules.

1. What is MSG? How is it produced?
2. Name three microbes used to produce vitamins.
3. What kind of microbes are typically used to produce antibiotics?

Microbiological Assay

Microbiological assay is a highly specialized application of the fact that certain organisms lack certain specific synthetic power, *i.e.*, are auxotrophs. *Lactobacillus plantarum*, for example, is unable to synthesize nicotinic acid ("niacin"). We may furnish the organism with a medium that is complete and satisfactory in all other respects but if niacin is lacking, absolutely no growth occurs. (Humans are no better off; without niacin they die of pellagra). If a minute amount (say, 0.01 microgram) per milliliter of niacin is added to the medium for *L. plantarum*, some growth will occur. More growth will occur in the presence of more of the missing factor. Up to the point of satiation or acidification, growth bears a linear relationship to the amount of the specific growth factor added.

For example, to assay the nicotinic acid content of fresh green beans, we prepare a medium for *L. plantarum* that is complete in all respects except niacin. This we omit. We now prepare two series (A and B) of 10 sterile tubes each. Each tube receives 10 ml. of the niacin-deficient medium. To each tube in series A we add known and graded amounts of pure niacin. To each tube in series B we add graded amounts of bean extract, niacin content unknown. All tubes are now inoculated with carefully washed (niacin free!) cells of *Lactobacillus plantarum*. Accurate, photometric measurements are then made of the growths (turbidities) obtained in the cultures. If the medium contains glucase, titrations of acidity instead of turbidity may be used as a measure of growth. By comparing growths in series A and B it is possible to estimate closely the concentration of niacin in the green beans. This method of estimation of growth factors is spoken of as *microbiological assay*.

Although the basic principle of all microbiological assays is the same, there are other methods of measuring the growth (or other physiological) response. These affect the cultural methods used. A commonly used procedure is the measurement of carbondioxide produced by fermentation of sugar in the test medium. Yeast is routinely used in the microbiological assay of *thiamine* (vitamin B_1) by this method. In *pyridoxine* (vitamin B_6)

assays, the mold *Neurospora* is the test organism. after sufficient incubation the culture is steamed and the entire mycelium of *Neurospora* is removed from the culture medium dried and weighed. Dry weight is directly proportional to concentration of pyridoxine in the sample of material being assayed. Another assay procedure depends on the spherophast-producing power of the assayed substance.

Certain organisms lend themselves very well to such assay procedures. *Lactobacillus casei* and *L. arabinosus* are easy to cultivate, relatively hardy, harmless and wholly dependent upon several growth-factors including various amino acids, riboflavin, biotin, pantothenic acid and nicotinic acid. Other organisms may be used for assay for other substances, for example, *Streptococcus lactis* of folic acid. Ultraviolet-induced, synthetically deficient auxotrophs of molds, yeasts ad bacteria are extremely valuable in assay work.

Even though the basic principle of microbiological assay is easily understood, the technological details are often exceedingly complex and filled with pitfalls. Many obscure factors affect the test organisms, and they may also undergo mutation and other changes without notice. Mutational and other injuries may be held to a minimum by storage of the stock cultures in containers with liquid nitrogen at –196°C. (–321°F). Temperature, pH and presence or absence of air may be of critical importance. For example, under *aerobic* conditions, *Lactobacillus lactis* will *die* before it will grow without vitamin B_{12}! *Anaerobically*, it sneers at vitamin B_{12}! There are many other examples. We may smile, but knowledge of this and many other peculiarities is essential to successful assay procedures.

Industrial Spoilage

In contrast with the useful activities of bacteria, a word may be said of their destructive action. Several causes of industrial spoilage (*e.g.*, "diseases" of fermentations) have been mentioned in this chapter and in the chapters on soil, food and water bacteria. Species of *Micrococcus, Alkaligenes Flavobacterium, Serratia, Clostridium,* coliform organisms, yeasts and molds are common causes of spoilage.

Each type of product is attacked by certain species of micro-organisms that can metabolize the substance especially well. For example, spoilage of cellulosic products such as lumber, telephone poles; paper; sisal, jute and flax fibers; tobacco and cotton is brought about by cellulose decomposers such as molds, various species of *Clostridium, Cellulomonas, Cytophaga* and many other such organisms of the soil. Fermentable substances such as syrups and beverages are attacked by yeasts, lactobacilli, organisms of the coli-aerogenes group and various environmental bacteria including the Genus *Clostridium*. Spoilage of proteins such as meats, fish, milk and so on result from the action of proteolytic species such as *Pseudomonas, Bacillus, Proteus, Micrococcus, Clostridium* and many others. Petroleum hydrocarbons are attacked by certain soil bacteria, as already mentioned, and rubber insulation of vital communication wires is attacked by bacteria and eucaryotic fungi.

Lactobacilli and *leuconostoc* species have already been noted as particular villains in the acid-food, fermentation and distillery industries. Species of both can ruin fruit or vegetable juice or various industrial mashes (beer and wine) during processing. They produce a buttermilk flavour. Pasteur found *Lactobacillus* and *Leuconostoc* causing "diseases" of beers and wines. They are just as active today. Lactobacilli also discolour meats, especially producing greenish discolouration (oxidized porphyrins) or cured hams and sausages.

The slimy dextran or levan-forming species, such as *Leuconostoc mesenteroides* and *L. Dextranicum* and some lactobacilli and micrococci, produce slimy and ropy conditions in a great variety of human endeavours; sugar refineries, pickle brines, dairy products, ham-curing cellar, and the like. These organisms prefer acidified products such as partly fermented foods, mashes and citrus juice. Examples of several of these types of spoilage have been given in discussions of the various products.

Development of undesirable flavours in fatty products such as butter, especially rancidity, is due in great part to the formation of butyric acid as a result of lipolysis. It is caused by species of *Aspergillus* and other molds, *Pseudomonas* species and streptococci related to *Str. liquefaciens*. These difficulties do not arise when clean equipment, clean milk and proper precautions to avoid contamination are used.

Proteolytic organisms, such as *Str. liquesfaciens* are responsible for undesirable bitter flavours and early spoilage of cheeses and other protein products. Gas production is usually caused by coliforms and *Clostridium*; putrefaction or digestion by *Clostridium, Pseudomonas* and *Bacillus*. Such conditions result mainly from dirty milk or other food equipment, or careless handling.

Included among other sabotage activities of bacteria are corrosion of the inside of structural aluminum alloys used in fuel tanks of jet-fuel aircraft. Pitting and scaling of the metal occurs beneath heavy, slimy growths of hydrocarbon-utilizing bacteria such as species of *Pseudomonas* and *Desulphovibrio* (see sulphate reduction). Some species of molds are also involved. Bacteria of the Genera *Mycobaterium* and *Nocardia,* among others, are also implicated in the deterioration of bituminous products, including asphalt highways and asphalt coatings and pipe-linings. The micro–organisms seem to utilize the high viscosity hydrocarbons and resins in asphalt.

Prevention of Spoilage

This, in each instance, is a problem that can be solved only by careful examinations of the process involved to find: (a) the nature of the organism(s) involved; (b) where the contamination is getting in, and (c) then devising means of excluding it. It is impossible to lay down a blanket rule for industrial spoilage in general. Everything depends on maintaining conditions unfavourable to, or excluding by asepsis, organisms that can grow on or in the particular product involved. This may involve complete steam sterilization of fermentation equipment (tanks, pipes, pumps); drying; refrigeration; aeration; the use of inhibitory salt, sugar or acid concentrations' radiation with ultraviolet light; exposure to sunlight; treatment with substances such as creosote, sodium benzoate and the like. In some processes specific antibiotics may be used, as penicillin and tetracyclines in alcoholic fermentation of molasses.